Prevention of Coronary Heart Disease
From the Cholesterol Hypothesis to ω6/ω3 Balance

World Review of Nutrition and Dietetics

Vol. 96

KARGER

Prevention of Coronary Heart Disease

From the Cholesterol Hypothesis to ω6/ω3 Balance

Volume Editor
Harumi Okuyama   *Nagoya*

Authors

Harumi Okuyama   *Laboratory of Preventive Nutraceutical Sciences, Kinjo Gakuin University College of Pharmacy, Nagoya, Japan*

Yuko Ichikawa   *Department of Health Promotion and Preventive Medicine, Nagoya City University School of Medical Sciences, Nagoya, Japan*

Yueji Sun   *Department of Psychiatry, Dalian Medical University, Dalian, China*

Tomohito Hamazaki   *Institute of Natural Medicine, University of Toyama, Toyama, Japan*

W.E.M. Lands   *College Park, Md., USA*

92 figures, 89 in color and 23 tables, 2007

Basel · Freiburg · Paris · London · New York ·
Bangalore · Bangkok · Singapore · Tokyo · Sydney

Harumi Okuyama
Kinjo Gakuin University College of Pharmacy
Laboratory of Preventive Nutraceutical Sciences
2 1723 Omori, Moriyamaku, Nagoya 463-8521 (Japan)
Tel. +81 52 798 7479, Fax +81 52 798 0754
E-Mail okuyamah@kinjo-u.ac.jp

Library of Congress Cataloging-in-Publication Data

Prevention of coronary heart disease : from the cholesterol hypothesis to
[omega] 6/ [omega] 3 balance / volume editor, Harumi Okuyama ; authors,
Harumi Okuyama ... [et al.].
        p. ; cm. -- (World review of nutrition and dietetics, ISSN 0084-2230
; v. 96)
 Includes bibliographical references and index.
 ISBN-13: 978-3-8055-8179-0 (hard cover : alk. paper)
 ISBN-10: 3-8055-8179-3 (hard cover : alk. paper)
 1. Coronary heart disease--Prevention. 2. Coronary heart
disease--Nutritional aspects. 3. Omega-6/omega-3 fatty acid ratio. I.
Okuyama, Harumi. II. Series.
 [DNLM: 1. Coronary Disease. 2. Coronary Disease--prevention & control.
3. Fatty Acids. 4. Hypercholesterolemia. 5. Neoplasms. W1 WO898 v.96
2007 / WG 300 P94545 2007]
 RC685.C6P6667 2007
 616.1'2305--dc22   2006030664

©Copyright 2007 by S. Karger AG, P.O. Box, CH–4009 Basel (Switzerland)
www.karger.com
Printed in Switzerland on acid-free paper by Reinhardt Druck, Basel
ISSN 0084–2230
ISBN-10: 3-8055-8179-3
ISBN-13: 978-3-8055-8179-0

Contents

Abstract

Prevention of Coronary Heart Disease
From the Cholesterol Hypothesis to ω6/ω3 Balance

The cholesterol hypothesis established half a century ago proposed a health benefit in (1) reducing the intake of saturated FAs (S); (2) reducing that of cholesterol, and (3) increasing that of polyunsaturated FA (P) that was essentially linoleic acid (LA) in vegetable oils, to lower serum cholesterol (TC; a TC value of 1 mmol/l corresponds to 38.61 mg/dl) and thereby reduce atherosclerosis-related diseases. Following epidemiological studies on Greenland natives and Danes in the 1970s, basic and clinical studies have gradually convinced people to accept the effectiveness of seafood ω3 FAs for the prevention of coronary heart disease (CHD) so that (4) 'increasing the intake of ω3 FAs' is now additionally recommended in some countries. Nowadays, these four recommendations are widely adopted in the medical fields in many countries, although the amounts of seafood intake differ much among countries and populations. However, the cholesterol-lowering activity of high-LA vegetable oils was transient, and diet recommendations (1–3) were essentially ineffective in reducing TC values in the long run (>several years). We propose that feedback control mechanisms in the body cause traditional cholesterol biomarkers to have different importance in interventions of short-term and long-term duration. More importantly, the association of high TC values with high CHD mortality differs several-fold in some populations with no significant differences among other populations. The proportion of familial hypercholesterolemia (FH) in high TC groups seems to be a critical factor in the reported high CHD mortalities. High CHD mortality in high TC groups may simply reflect the incidence and severity of FH cases. Because of inborn disorders of LDL receptors and their metabolism, FH cases develop hypercholesterolemia at younger ages, develop CHD at >10 times higher rates, and die younger than in non-FH cases. As a result, lipid nutrition and pathology of CHD in FH cases should be studied separately from those for the majority of non-FH cases in most populations. An important observation in recent studies is that higher TC values associate with lower cancer and all-cause mortalities in general populations and aged populations in which the relative proportions of FH are likely to be low. Although the effectiveness of statins, cholesterol synthesis inhibitors, in preventing CHD has been accepted, the benefits of statins may be mediated not by lowering TC and LDL-C, but rather by suppressing the formation of isoprenoid intermediates that have diverse activities. Furthermore, increased cancer and all-cause mortalities occurred in a Japanese intervention trial when a statin treatment lowered TC values from 264 to below 200 mg/dl. No benefit seems to come from efforts to limit dietary cholesterol intake or to lower TC values below approximately 260 mg/dl among general populations and aged populations ($\geq$40 years old in Japan and $\geq$50 years old in Austria). The cholesterol hypothesis regarded TC to be a mediator of disease that is increased by eating saturated fats and decreased by eating polyunsaturated fats. However, high TC can also be regarded as a biomarker for excessive mevalonate formation that accompanies excessive intakes of food calories which up-regulate gene expressions related to the levels of isoprenoid intermediates, cellular proliferation, carcinogenesis, inflammatory signaling mechanisms and unsuitable energy metabolism (chap. 1.4.). Biomarkers linearly linked to fatal CHD events need to be used in clinical epidemiology. In addition, an unbalanced intake of ω6 over ω3 polyunsaturated fats favors production of potent hormone-like eicosanoids whose actions lead to inflammatory and thrombotic lipid mediators and altered cellular signaling and gene expression which are major risk factors for CHD, cancers and shorter longevity (chap. 6.9.). High intakes of the ω6 linoleic acid do not correlate linearly with mortality, but the resulting high proportions of ω6 arachidonic acid in the highly unsaturated FAs of tissue lipids do correlate linearly with observed CHD fatalities. The health risk from high intakes of calories and saturated fats seems overcome by higher intakes of ω3 FAs and lower ω6/ω3 ratios in the diet. These favorable features are seen among Greenland and Mediterranean populations. Based on the data reviewed in this book, we recommend new directions of lipid nutrition for the primary and secondary prevention of CHD, cancer and all-cause deaths.

Chapter 1

The Cholesterol Hypothesis – Its Basis and Its Faults

1.1.

Cholesterol Hypothesis, Keys' Equation and Key Concepts for New Interpretations

Table 1

Cholesterol hypothesis, Keys' equation and key concepts

Observations and questions on the relationship of dietary lipids, TC and CHD
• Cholesterol crystal deposits are found in atherosclerotic plaques *Yes, but whether they are the cause or effect (consequence) of disease is not known.*
• Positive correlations are noted between the TC and CHD events *Yes, in some populations, but not all; CHD events differ approx. 4- to 8-fold at the same TC value*
• High LDL-C is often associated with more severe vascular disease *Yes, but harmful events associated with forming LDL from VLDL are seldom interpreted.*
• Uptake of oxidized LDL by macrophages is an early event of atherogenesis *Yes, but inflammatory processes that increase oxidized LDL are critical events in disease progression*
• LDL/HDL balance is better than TC as a predictor of atherosclerosis *TC and LDL-C are highly positively correlated, but the causal relationship between them and CHD is unclear. TC is probably a measure of CHD risk in groups with relatively large proportions of familial hypercholesterolemia (FH)*
Keys' empirical equation: $\Delta TC = 1.35 \times (2\Delta S - \Delta P) + 1.5\Delta Z$,
where ΔTC is the change in serum total cholesterol when ΔS and ΔP are changes in % of energy contributed by saturated and polyunsaturated fat, respectively, and ΔZ is change of square root of dietary cholesterol in mg/4 MJ
Note: A sharp decrease in the amount of ordinary fats in the American diet, without any changes in the amount of calories, lowered the TC level. The fall was rapid in the first few days, but after a few weeks there was an approach to a new plateau [cited from Keys et al., 1957]
In Hegsted's equation:
$\Delta TC = 2.16 \times \Delta S - 1.65 \times \Delta P + 6.77 \times$ Ingested cholesterol/day $- 0.53$ [Hegsted et al., 1965]
Cholesterol hypothesis, 1950s
Increasing the intake of high-LA oil (P) and decreasing that of animal fats (S) and cholesterol can decrease TC and thereby reduce CHD mortality
Key concepts used in our interpretation of the available data (chap. 1.5.) are:
Duration of intervention; proportion of FH in the subject population; relation to cancer and all-cause deaths; statin action on isoprenoid formation

1.2.

Current Nutritional Recommendations for the Prevention of Coronary Heart Disease Are Diverse among Scientists' Groups

Table 2

Dietary recommendations adopted by different groups

	S	P	Cholesterol
ATPIII of the NCEP[1]	S < 7 en%	P, up to 10 en% M, up to 20 en%	<200 mg/dl
MRFIT Study[2]	S < 8 en%	up to 10 en%	<250 mg/dl
AHA/ACC Guidelines[3]	S < 7 en%	P not indicated increased ω3 encouraged	<200 mg/dl
Jpn Soc Lipid Nutr[4,5]	not indicated	reduce ω6 from current 5–6 en%	not indicated
ISSFAL Workshop[6] (adequate intake for adults)	S < 8 en%	LA, 2~3 en%; ALA, 1 en%; EPA + DHA, 0.3 en%	not indicated

In these recommendations (1–3), the 'P' essentially stands for LA in vegetable oils and oil products.
[1] The Adult Treatment Panel III of the National Cholesterol Education Program (NCEP) [2002].
[2] Multiple Risk Factor Intervention Trial Research Group [1982].
[3] AHA/ACC Guidelines for Prevention of Heart Attack and Death in Patients with Atherosclerotic Cardiovascular Disease [2001]. A statement for healthcare professional from the American Heart Association and the American College of Cardiologists [Smith et al., 2001].
[4] The Japan Society for Lipid Nutrition recommends reducing the intake of LA. Hamazaki and Okuyama [2003].
[5] Summary by the President of Japan Society for Lipid Nutrition [Okuyama, 1997]. For the prevention of cancer, allergic hyper-reactivity, thrombotic diseases and others, a ω6/ω3 ratio of 2 or below was recommended [Okuyama et al., 1996; Okuyama, 2001].
[6] Simopoulos et al. [1999].

Our comments: The ATPIII recommendations from the NCEP have been adopted in Japan (Ministry of Health & Welfare, Japan and Japan Medical Association). We are very curious why the ATP III is currently proposed without defining P in terms of ω6 and ω3 when the MRFIT study failed to find beneficial effects of dietary recommendations similar to those of ATP (fig. 9). If P denotes LA (ω6) in vegetable oil as was the case for the delineation of Keys' equation, ignoring ω3 balance could be very risky. The current intake of LA is in the order of 6–7% in the USA and Japan, and the Japan Society for Lipid Nutrition and ISSFAL Workshop recommended substantial reduction of LA intake. Considering the competitive aspects of ω6 and ω3 FAs at many metabolic steps in our body, we support the recommendations by the Japanese Society for Lipid Nutrition and the ISSFAL to reduce the intake of LA (ω6) substantially to gain more benefits from limited supplies of ω3 dietary fats.

1.3.

Effect of Dietary Cholesterol and P/S Ratio of Fatty Acids on Total Cholesterol – Short- and Long-Term Effects Differ

Fig. 1

Increasing cholesterol intake elevates TC – A short-term effect

Weggermans et al. [2001], reproduced with kind permission from the *American Journal of Clinical Nutrition.* Copyright© 2006. American Society for Nutrition.

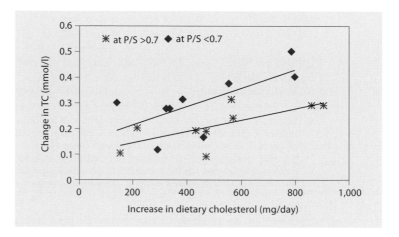

From 1974 to June 1999, 17 clinical studies (556 subjects) used crossover or parallel designs with control groups with 14 or more days of diets that differed only in the amount of dietary cholesterol or number of eggs. Adding 100 mg dietary cholesterol per day increased total cholesterol concentrations by 0.056 mmol/l (2.2 mg/dl) and HDL-cholesterol concentrations by 0.008 mmol/l (0.3 mg/dl).

Conclusion *(by the authors of the original paper): Dietary cholesterol raises the ratio of TC to HDL and, therefore, adversely affects the cholesterol levels associated with CHD. The advice to limit cholesterol intake by reducing the consumption of eggs and other cholesterol-rich foods may therefore still be valid.*

Our comments: These data involve changes obtained from a short-term intervention of several weeks. However, habitual intakes of large amounts of cholesterol during longer periods (fig. 2–4) do not correlate with higher TC levels.

Prevention of Coronary Heart Disease

Three groups of daily egg consumption in the Framingham Diet Study

Dawber et al. [1982], reproduced with kind permission from the *American Journal of Clinical Nutrition*. Copyright© 2006. American Society for Nutrition.

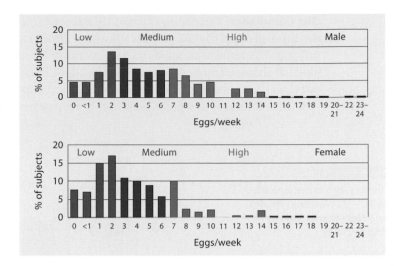

TC profiles in the three groups of egg intake in Framingham were similar – A long-term effect

Dawber TR, et al. [1982], reproduced with kind permission from the *American Journal of Clinical Nutrition*. Copyright© 2006. American Society for Nutrition.

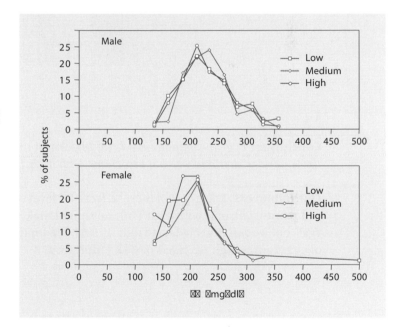

Conclusion from figures 2, 3: *Within the range of egg intake of this population differences in egg consumption were unrelated to blood cholesterol level or coronary heart disease incidence (relationships between TC and CHD are shown in figures 17 and 18).*

Our Comments: Despite significant differences in cholesterol intake among the three groups, the TC profiles were very similar to each other and no significant difference was noted in CHD events of the three groups. In epidemiological studies, Inuits ingested roughly 2-fold more cholesterol than Danes, but their TC values were slightly lower. Eggs are important sources of various vitamins, minerals and protein even in industrialized countries.

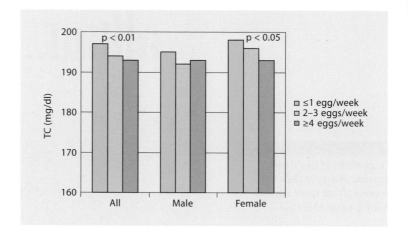

Fig. 4

Dietary egg is not linked to TC – US National Health and Nutritional Survey (1984–1994)

Data taken from Song and Kerver [2000].

Conclusion: *Egg consumption made an important nutritional contribution to the American diet and was not associated with high TC concentrations.*

Our comments: Dietary cholesterol affects TC level very little in the long term (>3 months). People with high TC levels might make daily efforts to eat fewer eggs, but we can find no data to indicate that habitual intake of a large number of eggs causes high TC levels or CHD mortality.

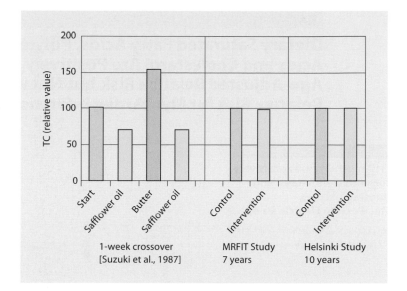

Fig. 5

Short-term and long-term effects of dietary FAs on TC are different

Data taken from Suzuki et al. [1970]; Multiple Risk Factor Intervention Trial Research Group [1982]; Strandberg et al. [1991]. See also, Keys et al. [1950].

One-week crossover study by Suzuki et al. [1984] (3 cases) and long-term intervention trials to raise P/S ratio and reduce cholesterol intake (7 years in MRFIT Study and 10 years in Helsinki Businessmen Study) are compared.

Conclusion: *Long-term dietary interventions did not alter TC while the effect was marked in a 1-week crossover study. Smoking rate decreased by 30% in the MRFIT Study, and mortality rates changed in the Helsinki Businessmen Study (fig. 10).*

Our comments: Raising P/S ratio of dietary FAs and/or reducing cholesterol intake are effective in lowering TC within a few days, but after a few weeks there is an approach to a new plateau from which Keys' equations were derived [Keys et al., 1957]. In the long term (several years), however, these dietary interventions are ineffective in substantially lowering TC. Serum lipoprotein profiles change shortly after dietary changes, but enzymes in our body adapt to new dietary conditions. Thus, the consequences of short- and long-term nutritional manipulations differ markedly.

One can argue that the ineffectiveness of long-term dietary interventions is due to progressively decreased compliance to the recommendations. However, intervention with dietary advice increased mortality for a subgroup of the MRFIT Study (fig. 9, below) and for the whole group in the Helsinki Businessmen Study (fig. 10; table 4), indicating that a lack of compliance is not likely to be a major cause for the failure to lower TC values.

The Cholesterol Hypothesis – Its Basis and Its Faults

7

1.4.

Dietary Saturated Fatty Acids, Polyunsaturated Fatty Acids and Cholesterol Are Positively Correlated with Age-Adjusted Relative Risk but Not with Multivariate Relative Risk for Myocardial Infarction

Fig. 6

CHD mortality was positively correlated with Keys' and Hegsted's scores – The Western Electric Study

Data taken from Shekelle RB et al. (1981) N Engl J Med 304:65–70.

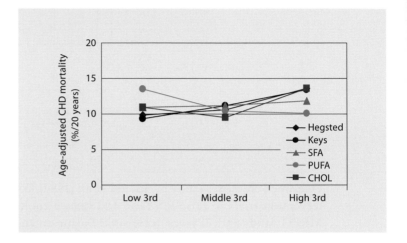

Diet and TC values were assessed for middle-aged male Americans (n = 1,900), and CHD mortality was followed for 20 years.

Conclusion: *Lipid composition of diet affects TC concentration and risk of coronary death in middle-aged American men.*

Our comments: Correlations between CHD mortality and intake of P (polyunsaturated FAs) or Keys' score were statistically significant, but the size of the observed effects on CHD was relatively small. Also, no statistical adjustment was made for dietary fiber intake, an important correlation factor noted in table 3.

Another factor to be considered is α-linolenic acid (ALA) present in vegetable oils. When the intake of vegetable oils (and LA) is relatively small, the intake of essential ALA might have protective effects on CHD, as pointed out by Hamazaki [2004].

Fig. 7

Dietary cholesterol co-varied with S intake, but correlated little with TC

Data taken from Health Professionals Follow-Up Study in the USA [Ascherio et al., 1996].

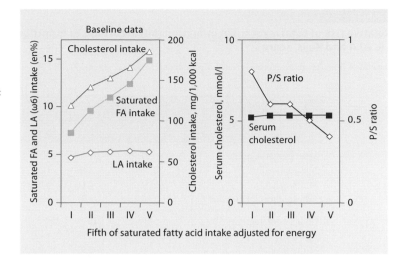

Baseline data

Fifth of saturated fatty acid intake adjusted for energy

Health professionals (45–75 years old, n = 43,757) were followed for 6 years from 1986.

Conclusion: *Given in the next data block.*

Our comments: Subjects hypercholesterolemic at baseline were excluded from the study of this population, and plasma TC values correlated very little with dietary cholesterol, saturated fat, or the overall P/S ratio of dietary FAs. In this sense, this subject population is quite different from that of the MRFIT Study in which those with upper 10–15% of TC values were selected.

Table 3

Relative risk of total myocardial infarction (MI) according to quintiles of dietary S, cholesterol, LA, ALA and Keys' score

Data taken from Health Professionals Follow-Up Study [Ascherio et al., 1996].

Factor and relative risk	Quintile					p value
	I	II	III	IV	V	
Saturated risk, g/day	17	21	24	27	33	
Age adjusted	1	1.16	1.05	1.21	1.44	0.002
Multivariate*	1	1.11	0.97	1.08	1.22	0.14
Adjusted for fiber intake**	1	1.01	0.84	0.90	0.96	0.69
Cholesterol, mg/day	189	246	290	338	422	
Age adjusted	1	0.96	1.12	1.14	1.34	0.003
Multivariate*	1	0.91	1.06	1.04	1.17	0.07
Adjusted for fiber intake**	1	0.86	0.98	0.94	1.03	0.48
Linoleic acid, g/day	7.6	9.6	11	12.6	15.4	
Age adjusted	1	1.24	1.12	1.13	1.08	0.89
Multivariate*	1	1.23	1.13	1.12	1.05	0.97
Adjusted for fiber intake	1	1.21	1.12	1.10	1.04	0.89
α-Linolenic acid, g/day	0.8	0.9	1.1	1.2	1.5	
Age adjusted	1	1.04	1.05	1.05	0.87	0.35
Multivariate*	1	1.01	0.99	1.00	0.82	0.10
Adjusted for fiber intake**	1	1	0.97	0.98	0.80	0.07
Trans unsaturated, g/day	1.5	2.2	2.7	3.3	4.3	
Age adjusted	1	1.24	1.33	1.40	1.57	0.0002
Multivariate*	1	1.20	1.24	1.27	1.40	0.01
Adjusted for fiber intake**	1	1.12	1.12	1.12	1.21	0.20
Keys' score	28	35	39.6	44.1	51.5	
Age adjusted	1	1.05	1.23	1.06	1.45	0.002
Multivariate*	1	1.02	1.14	0.95	1.23	0.15
Adjusted for fiber intake**	1	0.92	0.99	0.79	0.96	0.66

* Model includes age, body mass index, smoking habit, alcohol consumption, physical activity, history of hypertension, blood cholesterol, family history of MI before age 60, and profession.
** Additionally adjusted for fiber intake and adjusted for energy (continuous variable).

Conclusion: *These data do not confirm the strong association between intake of saturated fat and risk of CHD seen in international comparisons. However, they do fit the hypothesis that saturated fat and cholesterol intakes affect the risk of CHD as predicted by their effects on TC concentration. They also support a specific preventive effect of ALA intake.*

Our comments: Dietary S and cholesterol correlated weakly with MI, but the dietary effects on CHD mortality disappeared when adjusted for multiple factors, especially dietary fiber. Failure to reveal a causal relationship between LA and total MI, which we propose in Chapter 6, is due to the nutritional status of the average USA population in which the impact of dietary LA is near the saturating level (for comparison, see the competitive hyperbolic relationship shown in figure 70).

The harmful effect of dietary saturated FAs and PUFA on CHD was also questioned by Ravnskov [1998].

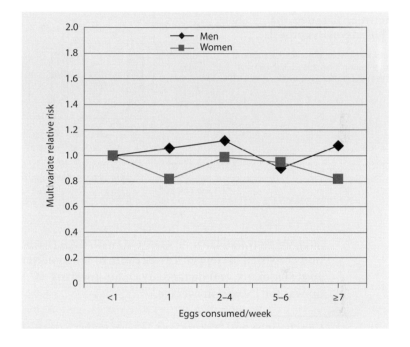

Fig. 8

No significant correlation between CHD risk and egg intake – A long-term prospective study in the USA

Data taken from Hu et al. [1999].

Men (40–75 years old, n = 37,851) were followed for 8 years and women (34–59 years old, n = 80,082) were followed for 14 years.

Conclusion: *Consumption of up to 1 egg/day is unlikely to have substantial overall impact on the risk of CHD or stroke among healthy men and women.*

Our comments: The great attention to possible harm from dietary cholesterol started before scientists showed that most TC comes from synthesis in the liver as excess food provides precursors while regulators of metabolism fail to limit the synthesis.

Increased Coronary Heart Disease Risk by Dietary Recommendations That Were Made Based on the Cholesterol Hypothesis

Fig. 9

A large-scale Multiple Risk Factor Intervention Trial (MRFIT) failed to prove the effectiveness of lipid nutrition based on the Cholesterol Hypothesis

Data taken from Multiple Risk Factor Intervention Trial Research Group [1982] JAMA 248:1465–1477.

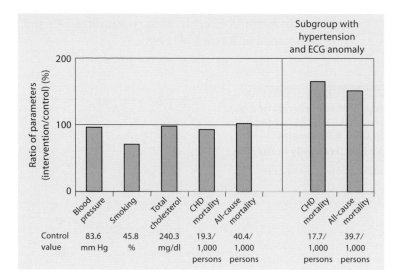

Subjects were male (35–57 years old) with high risk for CHD. Special intervention consisted of stepped-care treatment for hypertension, counseling for cigarette smoking, and dietary advice for lowering TC shown in table 2. Follow-up was for 7 years on average.

Conclusions and Discussion: *Three possible explanations for these findings are considered: (1) the overall intervention program, under these circumstances, does not affect CHD mortality; (2) the intervention used does affect CHD mortality, but the benefit was not observed in this trial of 7 years' average duration, with lower-than-expected mortality and with considerable risk factor change in the UC (usual care) group, and (3) measures to reduce cigarette smoking and to lower blood cholesterol levels may have reduced CHD mortality within subgroups of the SI (special intervention) cohort, with a possibly unfavorable response to antihypertensive drug therapy in certain but not all hypertensive subjects. This last possibility was considered most likely, needs further investigation, and lends support to some preventive measures while requiring reassessment of others.*

Our comments: Nutritional recommendations that were made based on the Cholesterol Hypothesis were found to be essentially ineffective in lowering TC and mortalities from CHD and all causes. In a subgroup with anomaly in ECG (electrocardiogram) and hypertension, the CHD and all-cause mortality rates were greater in the intervention group than in the control group. Hypotensive drugs were suspected as a cause for increased CHD events in this subgroup. However, CHD events among drug-treated patients were only 12% higher than average, which appears to be too small to account for the observed overall increase in CHD and all-cause mortalities following the multiple interventions. Perhaps imbalanced intakes of ω6 and ω3 played a role in the greater death rates, because analysis of dietary intakes (chap. 6) showed that CHD mortality in the 'usual care' group was negatively correlated with quintiles of dietary ω3 FAs (fig. 61), and analysis of serum phospholipid FAs confirmed that a 1 SD higher ω6 DGLA was associated with a 40% higher risk of CHD and a 1 SD higher ω3 DPA or DHA accompanied a 33% lower CHD risk [Simon et al., 1995].

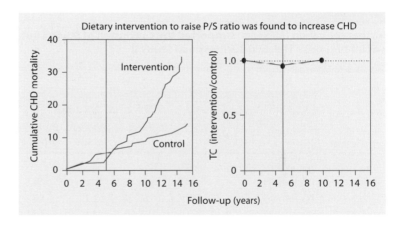

Those with high risk factors were divided into two groups; hypotensive and hypolipidemic drugs were used in the intervention group for the first 5 years and lifestyle intervention was continued for 15 years. The intervention included energy restriction, reduction of the intake of saturated fat, cholesterol, alcohol and sugar, and increase in the intake of P (mainly as soft margarine), fish, chicken, veal and vegetables. Increasing exercise was also advised and these advices were repeated.

Outcome: CHD and all-cause mortality were higher in the intervention group.

Conclusion: *The program reduced the incidence of stroke, but tended to increase that of coronary disease. In view of the small number of events, however, the significance of this observation remains uncertain. Results of Strandberg et al. [1991]: Multiple logistic regression analysis of treatments in the intervention group did not explain the 15-year excess cardiac mortality.*

Our comments: For the first 5 years, hypotensive and hypolipidemic drugs were used but the amount of drug users at year 10 was similarly low in the two groups (<30%), and the difference in CHD mortality between the two groups increased after 10 years, supporting the interpretation that the increased CHD mortality is due mainly to the dietary recommendations made throughout the 15 years. All-cause mortality was also significantly higher in the intervention group (1.4-fold).

Miettinen et al. [1972] reported earlier that CHD mortality decreased by raising the P/S ratio. However, that design was a 6-year crossover study, and its conclusion is contradicted by figure 10 where the difference in CHD mortalities became greater after 10 years of intervention while TC varied little.

Table 4

Further follow-up (28 years) supported the conclusion of Study I but the summary was ambiguous – Helsinki Businessmen Study II

Data taken from Strandberg et al. [1995].

Risk Factor	Group				
	Low risk n = 593	Control n = 610	Intervention n = 612	Excluded n = 563	Refused n = 867
Systolic Blood Pressure, mm Hg	130	135	135	144	136
TC, mmol/l	6.6	7.4	7.4	7.4	7.2
TG, mmol/l	1.2	1.7	1.6	1.7	1.5
Smokers, %	24.8	43.1	44.5	45.7	46.9
Mortality/1,000					
CHD	15.2	31.1	63.7	119.0	50.7
Neoplasm	28.7	47.5	44.1	53.3	55.4
Violent Death	25.3	1.6	26.1	21.3	18.5
Other	3.4	13.1	14.7	16.0	30.0
All-cause	79.3	106.6	155.2	259.3	179.9

Beside the two groups (control and intervention groups) listed in figure 10, three groups (low risk, excluded and refused groups) were added for analysis, and the follow-up period was extended.

Prevention of Coronary Heart Disease

Conclusion: *The traditional risk factors (smoking, blood pressure and cholesterol) are significantly associated with the 28-year mortality.*

Our Comments: By following up a period of 28 years, the authors reached the conclusion described above. However, the added groups differ significantly in the baseline data including the known risk factors of CHD as shown in this table, hence precise comparison is possible only between the control and intervention groups, the results of which supported the original conclusion (fig. 10).

It should be noted that increased LA intake (in the form of soft margarine in this intervention) was shown in these Helsinki Businessmen Study I and II to be associated with increased CHD and all-cause mortalities while suppressing the LA (arachidonic acid) cascade has been shown to be beneficial for the prevention of CHD (fig. 67; chap. 6). We are concerned that ethical problems can be raised for continuing the same intervention for 28 years despite the lack of observed benefit after the initial 15 years of follow-up (fig. 10).

Table 5

Low-fat, low-cholesterol diet in secondary prevention of CHD – An Australian Secondary Prevention Study

Data taken from Woodhill et al. [1978].

Group	P/S ratio of foods		Cholesterol intake, mg/day		TC, mg/dl		TG, mg/dl		Cumulative percentage survival[1]	
	P	F	P	F	P	F	P	F	P	F
Entry	0.5	0.5	486	508	272	281	186	189	100	100
After 5 years	0.8	1.7[b]	342	248[b]	262	250[a]	155	144	84	76[a]

[1] Survival data were calculated from the survival curves shown as a figure in the original paper.
[a] p <0.001; [b] p <0.001.

Four hundred and eight men with CHD participated in a trial of secondary prevention for 2–7 years in Australia. The group (F) with elevated P/S ratio of ingested FAs (p < 0.001), decreased cholesterol intake (p < 0.001) and lowered plasma TC levels (p < 0.01) exhibited a significantly smaller survival rate (p < 0.01) at 5 years of the trial.

Conclusion: *Multivariate analysis showed that none of the dietary factors were significantly related to survival. Prognosis was determined largely by the extent of coronary and myocardial disease as judged by the usual clinical parameters. Recreational physical activity had a strong favorable influence on survival when all other factors were kept constant.*

Our comment: Despite the conclusion drawn by the authors, we interpret the data to indicate that the intervention based on the 'Cholesterol Hypothesis' brought about unfavorable effects on CHD patients.

Dietary advice was revealed to be the most serious risk factor for CHD in Japan.

Fig. 11

An area-matched control study in relation to the Japan Lipid Intervention Trial (J-LIT)

Data taken from Yoshiike and Tanaka [2001].

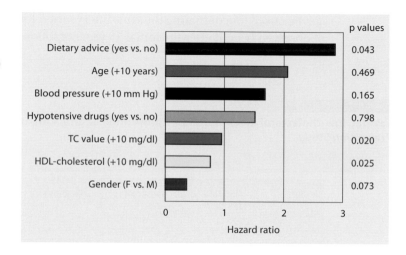

This is a kind of control study to the J-LIT in which the effects of a low dose of simvastatin were evaluated (fig. 42). Subjects with TC values of 220–299 mg/dl (n = 4,918, 35–70 years of age) and under 'usual care' were followed for 6 years. The incidence of MI and sudden cardiac death was 36 for 29,025 persons-year. Changes in TC value were –10.9 mg/dl for those without hypolipidemic drugs and –17.9 mg/dl for those who started taking hypolipidemic drugs.

Conclusions: *Dietary advice was strongly associated with high risk of CHD, but possible confounding factors must be considered (translated).*

Our comments: Currently, in the medical field in Japan, dietary advice is based on both the Cholesterol Hypothesis and the effectiveness of seafood ω3 FAs (Ministry of Health, Labor and Welfare Japan, and Japan Medical Society). That is, increasing the intake of P (LA in vegetable oils) and seafood ω3 FAs while decreasing the intake of saturated fats and cholesterol is recommended. However, LA (ω6) and other ω6 FAs compete with ω3 FAs at many enzymatic steps (chap. 7, section 2). Therefore, advising increased intake of both ω6 and ω3 FAs at the same time is not reasonable. We suggest that increased intake of LA may be a major cause for the observed increase in CHD incidence in the group with dietary advice. Higher intakes of LA accompanied higher rates of CHD (fig. 9–11; tables 4, 5), whereas decreasing LA intake was effective for the secondary prevention of CHD events (fig. 67). Similarly, the nutritional changes according to the Cholesterol Hypothesis were associated with increased CHD as seen in the Israeli Paradox [Yam et al., 1996], the Indian Paradox [Pella et al., 2003], and the Okinawa, Japan, story [Okuyama et al., 1996].

Summary

Cholesterol metabolism in our body is regulated in a complex manner, and the consequences of short-term and long-term dietary manipulations differ. Needless to say, dietary recommendations for the prevention of chronic diseases must be based on the results from long-term interventions or habitual nutrition (>several years). The difference, if any, in hypocholesterolemic activities of equal calorie amounts of animal fats and high-LA vegetable oils is very small in the long term. Because much TC is carried on LDL, TC values and LDL-C values are highly positively correlated, and either measure is associated with CHD in similar ways. The dietary recommendations based on the so-called Cholesterol Hypothesis actually increased CHD risk as noted in the MRFIT Study (subgroup), Helsinki Businessmen Study, Australian Study and Japanese J-LIT Area-Matched Control Study. Other studies also point out the increased CHD mortality by raised P/S ratio of foods (Israeli Paradox, Indian Paradox, Okinawan observation).

Chapter 2

Association of High Total Cholesterol with Coronary Heart Disease Mortality Differs among Subject Populations – Familial Hypercholesterolemia as a Key Concept

The association of TC with CHD (relative risk for the highest to lowest TC) differs among subject populations and among age groups. The mortality ratios range from more or less 5 (MRFIT Study, Tarui Report in Japan) to a nonsignificant level as in general populations in Japan and aged populations in the Western countries. The association is strong in younger generations (30–40 years of age) but very weak in aged groups (MRFIT Study, Framingham Heart Study). In some Japanese studies, the proportions of familial hypercholesterolemia (FH) were estimated or there were descriptions suggesting the presence of more than average proportion of FH (~0.2%). When subjects with high risks for CHD (e.g. high TC values) were selected, the subject populations are likely to include more than average proportions of FH.

So far, non-FH subjects with high TC values have been assumed to have pathological properties analogous to those of FH. However, there appear to be a number of differences in lipid metabolism in peripheral tissues (blood vessels) between FH and hypercholesterolemic non-FH cases. In this chapter, we put emphasis on the proportion of FH in the subject populations to interpret apparent discrepancies (variability) in relative risks of high TC in CHD mortality.

2.1.

Characteristics of Familial Hypercholesterolemia

Table 6

Characteristics of familial hypercholesterolemia

- Characterized by Brown & Goldstein as those with abnormal LDL receptor functions
- High TC and high LDL/HDL ratios even from young ages
- CHD events are higher by >10-fold even in heterozygote (Mabuchi et al., 1986)
- Lifespan is shorter by roughly 20% in heterozygote and 60% in homozygote
- Frequency is roughly 0.2% in the Western countries and in Japan (Fouchier et al., 2001)
- Relatively resistant to statin inhibition of cholesterol synthesis
- Up-regulation and inadequate feedback suppression of cholesterol synthesis

2.2.

Epidemiological Studies Led to Different Conclusions for the Relationship between Total Cholesterol and Coronary Heart Disease

Table 7

Association of high TC with high CHD differs among subject populations

Study	Subject	Mortality at high TC/that at low TC		
		CHD	cancer	all causes
USA				
MRFIT study	high risk group	4.8 at <40 years & 2.1 at 55–57 years	–	–
Seven countries study	general population?	2.3	–	–
Framingham study	general population followed for 30 years	<2 at <47 years n.s. at >48 years		0.3–2 (age-dependent)
Japan				
Tarul report	high risk group	5.6	–	–
Seven countries study	general population	n.s.	–	–
Yao Citizen	general population	n.s.	inverse correlation	inverse correlation
Hukul citizen	general population	n.s.	inverse correlation	inverse correlation
Ibaragi citizen	general population	1.6	inverse correlation	inverse correlation

n.s. = not significant.

Our comments: The association of TC with CHD mortality (mortality at the highest TC/that at the lowest TC) differs from very low (Framingham population followed for 30 years) to 2.3 (Seven Countries Study) and to ~5 (MRFIT subjects, 30 years of age) in the USA. Similarly in Japan, the association varies from insignificant levels (general populations) to 5.6 (Tarui Report, a government-supported survey). We found that the study groups that include or are suggested to include unusually high proportions of FH exhibit a stronger association of TC with CHD mortality than do general populations that include relatively small proportions of FH (~0.2%).

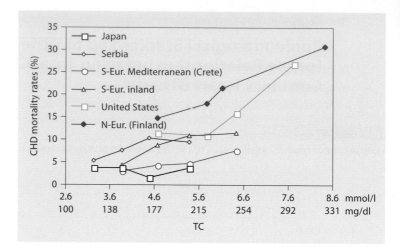

TC value was estimated in 1958–1964 for men aged 40–59 years (n = 12,467), and the CHD mortality rates per baseline cholesterol quartile were followed during the next 25 years.

Conclusion: *Across cultures, plasma cholesterol (TC) is linearly related to CHD mortality, and relatively higher CHD mortality rates are associated with higher cholesterol levels. The large difference in absolute CHD mortality rates at a given cholesterol level, however, indicates that other factors, such as diet, that are typical for cultures with a low CHD risk are also important with respect to primary prevention.*

Our comments: The association of TC with CHD mortality is very small in Japan and Greece (Crete) where people ingest relatively large amounts of seafood ω3 FAs [Marangoni and Galli, 2000]. It should be noted that the CHD mortality at the lowest TC group (190 mg/dl) in Finland was 4 times higher than that of the highest TC group (210 mg/dl) in Japan, indicating that factors other than TC are more critical in predicting mortality. The proposed beneficial effect of red wine was not supported by a clinical study performed in Denmark (table 15). Now, many lines of evidence support the conclusion that a major risk factor for CHD is 'high ω6/ω3 ratio of dietary FAs' (chap. 6). The relative risk of high TC (the slope in this figure) reflects the calculated ω6 HUFA% (a marker of ω6 eicosanoid tone) of serum lipids, which are 40% (Japan), 50% (Crete), 65% (South Europe), 70% (Serbia), 77% (USA) and 80% (Finland) (chap. 8).

Prevention of Coronary Heart Disease

2.3.

Data Rearranged to Convince Japanese People that High Total Cholesterol Is a Major Risk Factor for Coronary Heart Disease

Fig. 13

Apparent similarity in the TC and CHD relationship among MRFIT data (USA) and Japanese studies

Adapted from the Japan Atherosclerosis Society Guidelines for Diagnosis and Treatment of Atherosclerotic Cardiovascular Diseases, 1997.

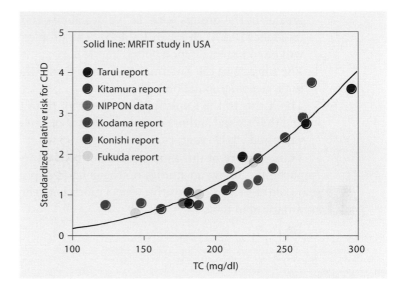

Conclusion: *These rearranged Japanese data fit the expected typical relationship showing a higher CHD mortality at TC values above ~220 mg/dl, a boundary value taken from the MRFIT Study. Thus, TC values above 220 or 240 mg/dl were interpreted to be risk factors for CHD (Guidelines from the Japan Atherosclerosis Society, 1997 and 2002).*

Our comments: This figure is often used in the Japanese medical field to convince people that high TC is the major risk factor for CHD. However, the data shown in this rearranged way is misleading, and one set of data is plotted incorrectly as pointed out below.

1 The association of TC with CHD events in the MRFIT Study is age-dependent; it is stronger in younger generations than in aged populations (fig. 16) as in the Framingham Study (fig. 17). The MRFIT data (solid line in this figure) appear to represent the data for a subgroup aged 40–44 years old in the MRFIT study. Furthermore, the subjects of the MRFIT study were selected for high risks of CHD (upper 10–15% of TC values), and are likely to include more than the average proportion of FH.

2 For Japanese data, the relative risk at a TC value of ~200 mg/dl is set at 1 for normalization, but the absolute incidence of CHD in Japan is 1/4–1/5 that in the USA.

3 The Tarui report described an analysis of complications among hyperlipidemic subjects (TC >220 mg/dl and/or TG >150 mg/dl), and not the results of the overall follow-up study. The proportion of FH in this population is approximately 19-fold greater as a whole, and >130-fold greater in the groups with TC values above 260 mg/dl than the average frequency (0.2%) in Japan (table 6). The problem of the NIPPON DATA will be commented upon below (fig. 15).

4 The subjects in the Kodama study were those who had A-bomb radiation, not a general population.

5 The X-axis in the Konishi report does not fit to this figure in dimension, and the original authors reported a lack of positive association between TC value and CHD event.

Thus, a figure of this kind should not be used in Japanese medical fields. Importantly, follow-up studies on the general populations in Japan (over 40 years old) revealed no positive associations between TC and CHD as partly summarized in table 7.

Fig. 14

Tarui Report clearly demonstrating the so-called bad and good cholesterol

Adapted from a report of research supported by the Ministry of Health and Welfare Japan that was summarized by Tarui [1987].

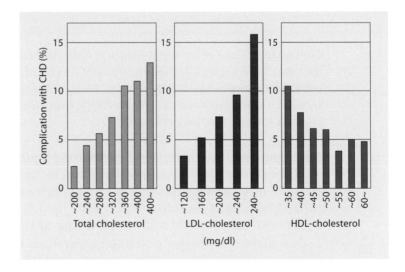

Prevention of Coronary Heart Disease

Subjects with high TC (>220 mg/dl) or high TG (>150 mg/dl) were selected (n = 10,313) and their complications with CHD were defined. TC and LDL-C were highly positively associated and HDL-C was negatively associated with CHD. The results were consistent with the general understanding that LDL delivers cholesterol to peripheral tissues and high LDL-C is a risk for CHD while HDL retro-transports cholesterol from peripheral tissue to liver and high HDL-C is a protective factor. The incidence of CHD was higher in FH (22.2% in men and 14.7% in women) than in non-FH type II hyperlipidemia (6.4% in men and 9.1% in women).

Conclusion *(a part of the Summary): The complication rate with ischemic heart disease (IHD) was highly positively associated with TC, and that of cerebrovascular disease was also associated with TC to a lesser extent. The association was more prominent when LDL was a marker. Complication rate with IHD was higher in lower HDL-C groups, suggesting the anti-atherogenic activity of HDL. The complication rate with IHD was markedly higher in FH than in non-FH even when TC values were adjusted. Moreover, the complication rate in FH was not associated with TC, but rather with serum TG levels (translated).*

Our Comments: The TC of this population ranged to much higher values than that in average Japanese as this study group included 19 times more FH than in general populations (0.2%), and the proportion of FH was 27% of the subgroups with TC values over 260 mg/dl. Our interpretation is that the highly positive associations observed between TC value and CHD events may simply reflect the proportion and severity of FH in this group.

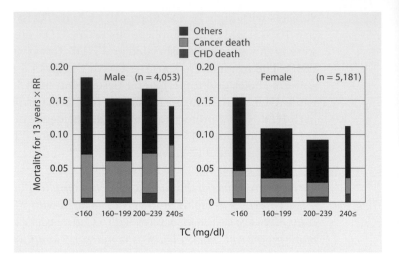

Fig. 15

1980 Survey on Circulatory Diseases – NIPPON DATA

Data taken from a follow-up study supported by the Ministry of Health and Welfare Japan [Ueshima, 1997; Okamura et al., 2003]. The width of the column is proportional to the number of subjects. In the original data RR at TC of 160–199 mg/dl was set at 1.

Results: Hypertension and smoking were risk factors for circulatory diseases in both the younger (below 74 years at the end of follow-up) and older (over 75 years) groups, but hypercholesterolemia was not a risk factor for circulatory diseases in both groups. However, hypercholesterolemia was one of risk factors for the mortality from ischemic heart occlusion in men of both groups.

Conclusions: *Similar to Western populations, it is recommended to provide screening for hypercholesterolemia in Japan, especially for males, although its attributable risk for coronary disease might be small.*

Our comments: The association of high TC with CHD events was relatively strong in this population. However, the presence of relatively large numbers of unusually hypercholesterolemic subjects was suggested in the original Japanese report because the median TC value was significantly lower than the mean TC value. The English version of the report describes this study group as a 'general population'. However, this was a survey performed before the start (1982) of the free medical check-up system supported by the Japanese government, hence this subject population is likely to differ from the general population, the follow-up study of which started after the 1982 law was in effect (table 7).

2.4.

The Association of Total Cholesterol with Coronary Heart Disease Differs among Different Age Groups, Which Is Likely to Be Correlated with Familial Hypercholesterolemia

Fig. 16

Age-specific adjusted relative risk of CHD by TC quintile in the MRFIT Study

Data taken from Kannel et al. [1986].

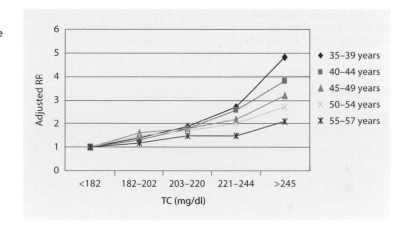

This randomized controlled trial involved men aged 35–57 years, who were free of overt evidence of CHD but at high risk of CHD because their levels of TC, blood pressure and cigarette smoking placed them in the upper 10–15%.

Conclusion: *The strength of the association of each of the risk factors with CHD and all-cause mortality rates diminished while deaths attributable to the risk factors increased because of the higher death rates in older men. It is estimated that elimination of these risk factors (TC, blood pressure, and smoking) has the potential for reducing the CHD mortality rate by two thirds in 35- to 45-year-old men, and by one half in 46- to 57-year-old men.*

Our comments: Because the subjects were selected for high TC levels (upper 10–15%), the MRFIT population is likely to include more than the average proportion of FH. A significant proportion of those who had developed hypercholesterolemia by 30 years of age are likely to have inborn disorders, and the relative proportion of FH was much higher in this group than in the general population. Thus, our hypothesis that high CHD mortality is mostly a reflection of the proportion and severity of FH, particularly in younger generations, also applies to this case. It should be noted that the relative risk of high TC for CHD is calculated from the next data (fig. 17) to be 1.1 (aged group) to 1.6 (younger group) in the Framingham Study.

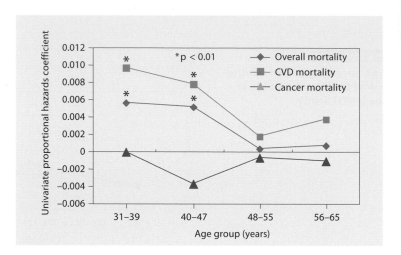

Fig. 17

Cholesterol and mortality – 30 years of follow-up from the Framingham Study: Association of TC with CHD events differs among different age groups

Data taken from Anderson et al. [1987]. Similar data in Neaton and Wentworth [1992].

From 1951 to 1955, TC levels were measured in 1,959 men and 2,415 women aged between 31 and 65 years who were free of CVD (CHD) and cancer, and the cause-specific mortality was followed for the subsequent 30 years. In the youngest two age groups stratified by age group and sex, β is estimated to be 0.0091 (p < 0.01), which corresponds to a 9% greater CVD (CHD) death rate for each 10 mg/dl change in TC values. The ratio of mortality at 240 mg/dl to that at 180 mg/dl is calculated to be 1.54 in these younger groups. Association between TC and all-cause mortality was significant in the younger two age groups but not in the older two age groups.

Conclusion: *This study and the Coronary Drug Project results on nicotinic acid therapy show a direct association of cholesterol levels with lengthy follow-up. This suggests that evidence linking cholesterol reduction with lower overall mortality in Lipid Research Clinics Coronary Primary Prevention Trial may become stronger given further follow-up.*

We believe those who would argue that low serum cholesterol levels should be avoided – because they pose a higher risk of death from cancer or other causes – cannot support such a stand when the pattern of mortality during a 30-year follow-up period is considered. We further believe values for TC are a useful screening device for considering alteration of lipids in individuals young-

Prevention of Coronary Heart Disease

er than 50 years. For individuals older than 50 years, lipoprotein fractions and apoprotein values are likely to be more appropriate screening devices than measurements of TC.

Our comments: Correlation between TC values and CHD mortality was clear in the younger generations but not in the generations over around 50 years old. Those with high TC values at 30 and 40 years of age are likely to include groups with genetic disorders such as familial hypercholesterolemia (FH), who develop CHD at >10 times higher rates and die younger (fig. 20). On the other hand, the results of an 18-year follow-up of Framingham men were reviewed [Kannel et al., 1979], and the mortality ratio at the highest to lowest TC was relatively high (3–5). However, the high ratio was seen only for men aged 35 years, and high TC was not a predictor of CHD risk beyond age 65.

The difference between the Framingham Study and MRFIT Study is striking in terms of relative risk. It should be noted that subjects in the Framingham Study are general populations while hypercholesterolemic subjects were selected in the MRFIT Study, and the proportion of FH is very likely to be greater in the MRFIT Study subjects.

It should also be noted that in Japan the CHD mortality is roughly 1/4 the level of the USA and the effect of TC on cancer mortality is likely to have been evaluated more accurately than in the USA because cancer is the leading cause of death in Japan. In several Japanese studies on general populations, the 'higher cancer mortality in the lower TC group' has been revealed (chap. 3).

Fig. 18

Crude mortality within age categories by sex and TC interval – Based on Framingham data

Data taken from Kronmal et al. [1993].

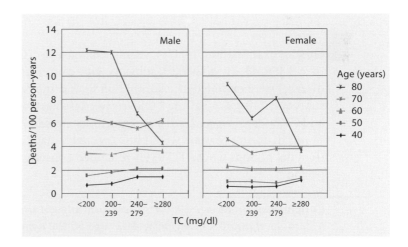

Conclusion: *TC levels and mortality risk are shown as a function of age. At present, there is no definitive basis for recommending lipid-lowering treatment in elderly men and women.*

Our comments: In the aged groups, low TC was a predictor of shorter survival. In the younger age groups, slightly higher mortality was observed in the high TC group, which is likely to reflect the characteristics of FH (table 6).

Fig. 19

Changes in TC during 14 years and subsequent mortality during 18 years – Framingham Study

Data taken from Anderson et al. [1987].

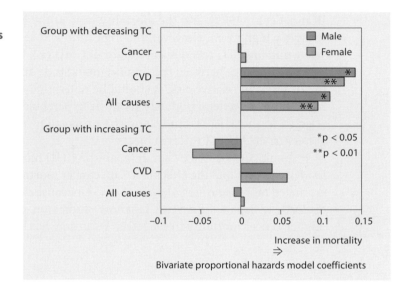

During examination 1–7 (14 years), changes in TC values were evaluated (n = 3,456, 43–79 years old), subjects with cancer and CVD (cardiovascular disease) were excluded at examination 8, and the rest were followed for the subsequent 18 years. Falling TC levels were associated with increased all-cause and CVD (cardiovascular disease) mortality. Persons whose TC levels dropped 14 mg/dl during 14 years would be expected to have an 11% higher death rate than the persons whose TC levels remained constant or rose during the same period.

Conclusion: *Under age 50 years these data suggest that having a very low cholesterol level improves longevity. After age 50 the association of mortality with cholesterol values is confounded by people whose cholesterol levels are falling – perhaps due to diseases predisposing to death.*

Our comments: These observations were apparently inconsistent with the 'Cholesterol Hypothesis', and have not been evaluated adequately in the medical fields worldwide. Now, many lines of evidence from different countries indicate that the higher the TC the lower are the cancer and all-cause mortality rates among general populations as will be shown in chapter 3.

2.5.

Lipid Metabolism Is Likely to Differ Qualitatively between Familial Hypercholesterolemia and Hypercholesterolemic Non-Familial Hypercholesterolemia Cases, and the Observations in Familial Hypercholesterolemia Cases Should Not Be Extended Directly to Non-Familial Hypercholesterolemia

Fig. 20

Relative proportions of FH and non-FH are likely to differ among age groups and among selected populations – A schematic model and a new hypothesis

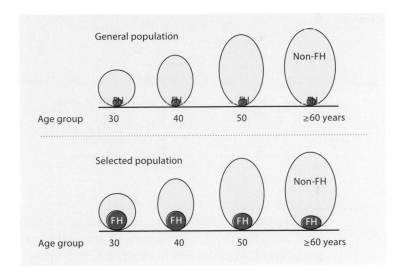

Our comments: The proportion of FH would not change much within centuries but TC levels change within decades due to the changes in lifestyle. TC levels increase with aging (up to certain ages), even in non-FH cases. The number of FH decreases with aging because of shorter survival, and the relative proportion of FH in the high TC group decreases with aging. Subject populations selected for high TC levels (MRFIT Study, Tarui Report) are likely to include greater proportions of FH than in general populations.

Our interpretation is that high TC is not a mediator causing CHD, and high CHD mortality in the high TC groups reflects the high proportion of FH in those groups. This interpretation (new hypothesis) is supported by the fact that no positive associations were observed between TC and CHD among the Japanese general populations over 40 years of age (fig. 30–35), the Austrian general populations over 50 years of age (fig. 26–29), the Framingham general populations over 48 years of age (fig. 17, 18), Japanese-Americans over 71 years of age (table 9; fig. 25), and a Netherlands' population over 85 years of age (fig. 24).

Fig. 21

FH and non-FH are likely to differ qualitatively in the cholesterol and energy metabolism in peripheral tissues

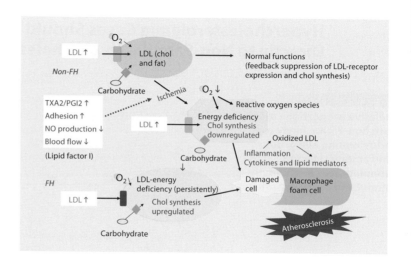

Our Explanation: Plasma LDL supplies FAs (a source of energy and essential FAs) and cholesterol to peripheral tissues and blood vessels. Normal cells in peripheral tissues respond to elevated LDL availability by reducing the number of LDL receptor and by feedback suppressing the expression of enzymes for intracellular cholesterol synthesis. Occasionally, cells encounter ischemia, and the supply of oxygen to cells is reduced. As a result, reactive oxygen species (ROS) produced from ischemic mitochondria and inflammatory cells oxidize LDL and injure cells. The damaged cells are digested by macrophages which form foam cells, a long-recognized process in atherogenesis.

In peripheral cells of FH cases, LDL is not transported into cells effectively, and the supply of cholesterol, FA energy and essential FAs through this path-

         Prevention of Coronary Heart Disease

way is limited persistently. Cholesterol synthesis is upregulated and the levels of isoprenoid metabolic intermediates are likely to be elevated and to activate oncogene products (Ras and Rho). Persistent limitation of oxygen and energy accompanied by excessive LDL FAs in peripheral cells would damage cells and lead to atherogenesis and shorter longevity.

A rabbit model is often used to study atherosclerosis, and fatty streaks are observed in blood vessels due to lipid deposit above the endothelial cell (EC) layer, unlike the lipid deposit observed below the EC layer in human atherosclerosis. Because rabbits adapted evolutionarily to foods containing limited amounts of cholesterol, regulation of the number of LDL receptor and other metabolic systems do not meet the supply of large amounts of cholesterol when used as an atherogenic model. In this sense, the Watanabe inheritable rabbit is considered a model for human FH cases.

Although the above interpretation is still uncertain, we find no evidence to assume that FH and hypercholesterolemic non-FH are the same in the mechanisms of developing CHD. FH and non-FH must be analyzed separately.

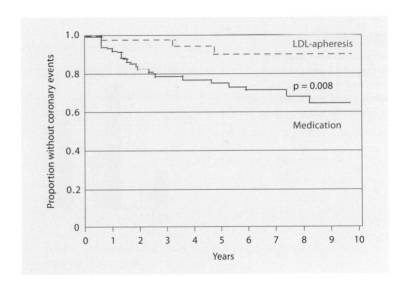

Fig. 22

Long-term efficacy of low-density lipoprotein apheresis on coronary heart disease in familial hypercholesterolemia

A part of the original figures was reproduced from Hokuriku-FH-LDL-Apheresis Study Group [Mabuchi et al., 1998], with kind permission from Elsevier.

This study describes the long-term (6 years) safety and efficacy of intensive cholesterol-lowering therapies with low-density lipoprotein (LDL) apheresis in heterozygous FH patients with CHD. One hundred and thirty heterozygous FH patients with CHD documented by coronary angiography had been treated by cholesterol-lowering drug therapy alone (n = 87) or LDL aphaeresis combined with cholesterol-lowering drugs (n = 43). Serum lipid levels and outcomes in each treatment group were compared after approximately 6 years. Both treatment groups had significant reductions in TC, LDL-C, and HDL-C levels. LDL aphaeresis significantly reduced LDL-C from 7.42 to 3.13 mmol/l (58%) compared with the group taking drug therapy, from 6.03 to 4.32 mmol/l (28%). With Kaplan-Meier analyses of the coronary events including nonfatal myocardial infarction, percutaneous transluminal coronary angioplasty, coronary artery bypass grafting, and death from CHD.

Conclusion: *The rate of total coronary events was 72% lower in the LDL-apheresis group (10%) than in the drug therapy group (36%) (p = 0.0088). It is concluded that LDL-apheresis is effective as treatment of CHD in FH heterozygote, and may become the therapy of choice in severe types of FH.*

Our Comments: Apheresis was shown to be very useful for the protection of coronary events in FH. It removes LDL and HDL as well as PAF mimics (peroxidized phospholipids) that activate platelet aggregation.

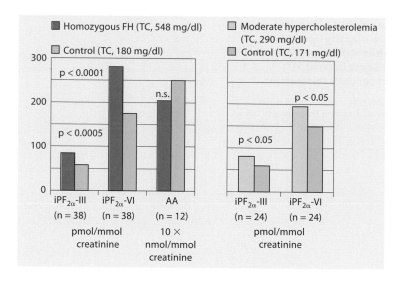

Fig. 23

Increased formation of distinct F2 isoprostanes (free radical catalyzed products of arachidonic acid) in hypercholesterolemia

Data taken from Reilly et al. [1998].

Urinary excretion of the 2 F2 isoprostanes was significantly increased in hypercholesterolemic patients, whereas substrate AA in urine did not differ between the groups. iPF2alpha-III was elevated ($p < 0.0005$) in homozygous familial hypercholesterolemic patients (n = 38) compared with age- and sex-matched normocholesterolemic control subjects (n = 38), as were levels of

Prevention of Coronary Heart Disease

iPF2alpha-VI (p < 0.0005). Serum cholesterol correlated with urinary iPF2alpha-III (r = 0.41; p < 0.02) and iPF2alpha-VI (r – 0. 39; p < 0.03) in HFH patients. Urinary excretion of iPF2alpha-III (p < 0.05) and iPF2alpha-VI (p < 0.05) was also increased in moderately hypercholesterolemic subjects (n = 24) compared with their controls. Urinary excretion of iPF2alpha-III and iPF2alpha-VI was correlated (r = 0.57; p < 0.0001; n = 106). LDL iPF2alpha-III levels (ng/mg arachidonate) were elevated (p < 0.01) in HFH patients compared with controls. The concentrations of iPF2-III in LDL and urine were significantly correlated (r = 0.42; p < 0.05) in HFH patients.

Conclusion: *Asymptomatic patients with moderate and severe hypercholesterolemia have evidence of oxidant stress in vivo.*

Our comments: FH subjects may have continual exposure of vessel walls to high amounts of TG and NEFA that promote increased inflammatory events and oxidant stress. The continual oxidant stress contrasts with the intermittent transient postprandial stress in non-FH people (see more in chap. 8). The greater oxidant stress in homozygous FH patients is evident from elevated excretion in urine of distinct isoprostanes as well as their presence in oxidized LDL. Selective removal of LDL lipoprotein and its oxidized phospholipid irritants by therapeutic apheresis lowered inflammatory markers and significantly improved electrocardiographic and angiographic measures indicating a reduction of progression of coronary lesions [Bosch et al., 2004]. Even a single LDL apheresis improved the coronary artery diameter and blood flow response to acetylcholine [Igaraschi et al., 2004]. This result indicates an improved vascular endothelial response of vasodilatory nitric oxide production (discussed further in chap. 4 and 8) once the toxic overload was removed. Chapter 8 examines the role of oxidants in impairing endothelial formation of nitric oxide.

Summary

The impact of TC on CHD (the mortality at the highest TC/lowest mortality in the case of U-shape profile or the relative risk of the highest to lowest TC) varies from nonsignificant levels to more or less 5 depending on the subject populations. From Japanese studies in which the proportions of FH in the subjects were described or suggested, we noticed that the relative risk is high whenever the proportion of FH is estimated to be high in the subject populations.

Age-dependent variations of the relative risk (MRFIT Study, Framingham Study and Austrian VHM&PP Study) could be interpreted rationally, assuming that high TC groups in younger age include more FH than those of older ages. Similarly, hospital-based surveys tend to include more FH than general populations; MRFIT Study and Framingham Study performed in the USA are essentially different in this context because the former subjects were selected for possible risk factors including upper 10–15% of TC values.

We emphasize that FH and hypercholesterolemic non-FH are metabolically different, and that high TC is not associated with CHD in general populations in which the proportions of FH are relatively small.

Chapter 3

Cancer and All-Cause Mortalities Are Lower in the Higher Total Cholesterol Groups among General Populations

Formerly, hospital-based surveys supported by the government were more common in Japan, and the results of such studies were mainly reflected in the medical fields. Even in such studies, CHD mortality was very low and leading scientists in this field encountered much difficulty in harmonizing the medical care for CHD with that in the USA (fig. 13). Since 1982, when Japanese government started supporting general populations to receive medical examinations for free, their data on biochemical parameters have accumulated, and several follow-up studies were published. Results of a large-scale follow-up study performed on a general population of Korea were also published, which were more or less similar to those on Japanese general populations in that TC was associated relatively little with CHD mortality.

Looking for similar data in the Western countries, we found several reports which were not consistent with the classical 'Cholesterol Hypothesis', some of which were described in the previous chapter. Altogether, we reached a new conclusion that efforts to lower TC should not be made at least for general populations because low TC value was a predictor of high cancer and all-cause mortalities. This conclusion most probably applies not only to Japan and Korea but also to general populations over 50 years old in the Western countries.

3.1.

Reports Showing Longer Survival in the Higher Total Cholesterol Groups Except for the Highest Total Cholesterol Group in Western Countries

Table 8

No significant positive correlations were observed between TC and CHD mortality, but inverse associations were seen between TC and mortality from cancer and/or all causes in aged populations of the Western countries

Report and Reference	Subjects and Period	Correlation of TC with	
		cancer mortality	all-cause mortality
Women in France Forette et al. [1989]	n = 92 (F), >60 years 5 years	no data	reverse J shape[a]
Honolulu Heart Program Shatz et al. [2001][b]	n = 3,572 (M), 71–93 years 20 years	no data	inverse correlation or reverse J
Oldest old in the Netherlands Weverling-Rijnsburger et al. [1997]	n = 724, >85 years 10 years	inverse correlation	inverse correlation
EPESE Study in the USA Krumholz et al. [1994]	n = 997, >70 years 4 years	not significant	not significant
Framingham Study, USA [Anderson et al. [1987][c]	n = 1,959 (M), 31–65 years 30 years	inverse correlation[d]	inverse correlation[d]

[a] Events were less with higher TC level (inverse correlation), except for the highest TC group in which the event is higher than at the bottom level.
[b] Japanese American. Note that HDL and LDL data do NOT fit as expected.
[c] Relative risk for the highest and lowest TC group was 1.6 (31–39 years old) and 1.1 (48–55 years old), which were much lower than in the MRFIT study; 4.9 (35–39 years old) and 2.1 (55–57 years old) (fig. 16).
[d] Decreasing TC during 14 years was associated with increasing mortalities from cancer and all causes during the subsequent 18 years (fig. 19).

Our Comments: TC values increase with aging to a certain age, and relative proportions of FH in the high-TC groups would be smaller in older populations (fig. 20), in which no significant positive correlations were seen between TC level and CHD event. Detailed data from the 'Oldest old in the Netherlands' are presented below (fig. 24). When the proportion of FH is likely to be relatively small, the higher the TC level the lower were the cancer and all-cause mortalities. This is in accordance with the observations on general populations (over 40 years old) in Japan as will be shown below (see also table 7).

Fig. 24

Ten years' follow-up of the oldest old in the Netherlands

A part of the original figure was reproduced from Weverling-Rijnsburger et al. [1997], with kind permission from Elsevier.

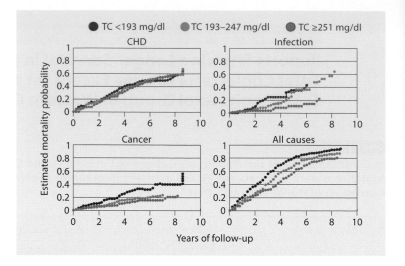

Those who are over 85 years old were followed for 10 years.

Conclusion: *In people older than 85 years, high TC concentrations are associated with longevity owing to lower mortality from cancer and infection. The effects of cholesterol-lowering therapy have yet to be assessed.*

Our comments: The proportion of FH is likely to be very low in older populations because of shorter survival, and the proportions of non-FH are likely to be greater in aging populations than in average (fig. 20).

Table 9

Cholesterol and all-cause mortality in elderly people from the Honolulu Heart Program: A cohort study

Data taken from Schatz et al. [2001].

TC group, mg/dl	149 (n = 904)	178 (n = 858)	199 (n = 902)	231 (n = 908)
Age-adjusted RR for all-cause mortality	1	0.72*	0.60*	0.65*

* p < 0.0001.

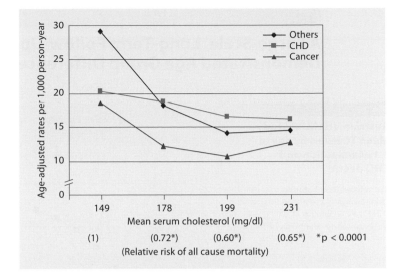

Fig. 25

Cholesterol and all-cause mortality in elderly people from the Honolulu Heart Program: A cohort study

Reproduced from Schatz et al. [2001], with kind permission from Elsevier.

Lipid and TC were measured in Japanese-American men (71–93 years old, n = 3,572), followed for 6 years, and the results were analyzed by a Cox proportional hazards model. Following analysis (fig. 25) revealed that mortalities from CHD, cancer and others were lower in the higher TC groups.

Conclusion: *These data cast doubt on the scientific justification for lowering cholesterol to very low concentrations (<4.65 mmol/l, 180 mg/dl) in elderly people.*

Our comments: TC value tended to decrease along with aging in this population. No scientific evidence was found to lower TC values in aged populations (similar observations in table 8). Again, it is clear that cholesterol-lowering medication is not applicable to these elderly people.

A Large-Scale, Long-Term Follow-Up Study in Austria Demonstrated Age Group Differences in Mortalities

Fig. 26

Austrian VHM&PP Study – Mean TC levels by age at 1st examination and CHD death

Reproduced from Vorarlberg Health Monitoring and Promotion Programme (VHM&PP) [Ulmer et al., 2004], with permission from Mary Ann Liebert, Inc.

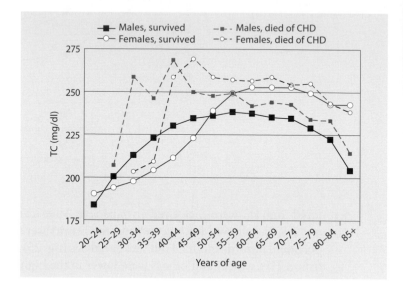

Men (n = 67,413) and women (n = 82,237) aged 20–95 years underwent 454,448 standardized examinations in the 15-year period 1985–1999. The role of high TC in predicting death from CHD could be confirmed in men of all ages and in women under the age of 50.

Conclusion: *See figures 27–29.*

Our comments: There were differences in the baseline TC values by ~50 mg/dl between the groups that died of CHD and the surviving subjects only in younger generations; men below 50 and women below 55 years of age. After about age 55, the differences in the TC values of the groups that died and the surviving groups were much less (<8 mg/dl) even in men. We interpret these data to indicate that groups with very high TC in younger ages include more proportions of FH which died young, hence the difference in TC values of the younger groups that died or survived was much greater than in aged groups in which relative proportions of FH were probably smaller (fig. 20). Such big differences were not observed between the groups that died from all causes and those which survived (fig. 27–29).

Fig. 27

Austrian VHM&PP Study – Mean TC levels by age at 1st examination and mortalities from CHD

Data taken from tables in Ulmer et al. [2004].

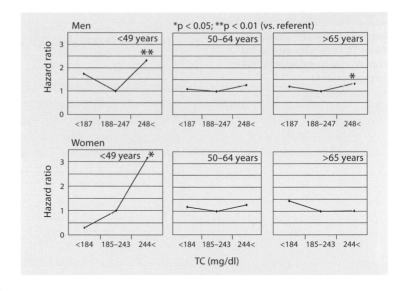

Fig. 28

Austrian VHM&PP Study – Mean TC levels by age at 1st examination and mortalities from cancer

Data taken from tables in Ulmer et al. [2004].

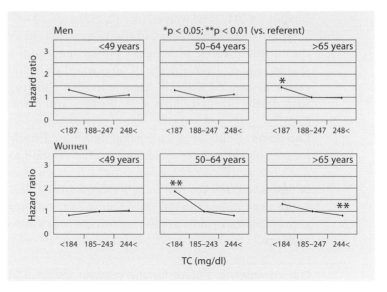

Fig. 29

Austrian VHM&PP Study – Mean TC levels by age at 1st examination and mortalities from all causes

Data taken from tables in Ulmer et al. [2004].

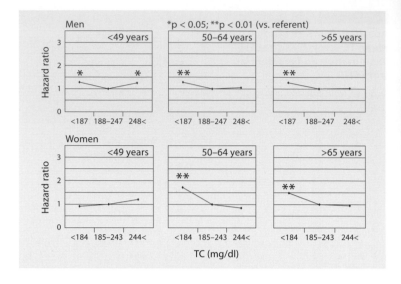

TC (mg/dl)

Men (n = 67,413) and women (n = 82,237) aged 20–95 years underwent 454,448 standardized examinations in the 15-year period 1985–1999. In men, across the entire age range, although of borderline significance under the age of 50, and in women from the age of 50 onward only, low cholesterol was significantly associated with all-cause mortality, showing significant associations with deaths through cancer, liver diseases and mental diseases.

Conclusion: *This large-scale population-based study clearly demonstrates the contrasting patterns of cholesterol level in relation to risk, particularly among those less well studied previously, that is, women of all ages and younger people of both sexes. For the first time, we demonstrate that the low cholesterol effect occurs even among younger respondents, contradicting the previous assessments among cohorts of older people that this is a proxy or marker for frailty occurring with age.*

Our comments: Except for the younger groups (20–49 years) in which the relative proportions of FH in the high TC groups are likely to be greater and the relationships between TC and mortalities exhibit complicate profiles, the association of high TC with CHD was relatively small after 50 years of age (fig. 27), and the mortalities from cancer (fig. 28) and all causes (fig. 29) were lower in the higher TC groups both in men and women. For the people older than 50 years in this population, cholesterol-lowering medication does not appear to be appropriate.

3.3.

Reports on Japanese General Populations Showing Lower Mortalities from Cancer and All Causes in the Higher Total Cholesterol Groups

Yao citizens (≥40 years old) with general medical examinations were followed for 8.5 years

Data taken from Iso et al. [1994].

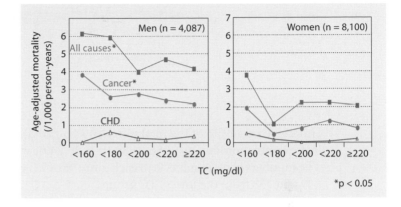

Baseline data were collected during 1975–1984 for citizens of Yao, Osaka (40–69 years of age, n = 12,187), and they were followed for 8.9 years on average up to 1988.

Conclusion: *There was a significant inverse association of TC with total and cancer mortality for men, and no significant association for women.*

Our comments: In a general population with a relatively small proportion of FH, no significant correlation was observed between TC and CHD. Importantly, cancer and all-cause mortalities were lower in the male groups with higher TC values. A similar tendency was noted for females although the association was not statistically significant.

Fig. 31

Fukui Citizens with general medical examinations were followed for 5 years

Data taken from Shirasaki [1997].

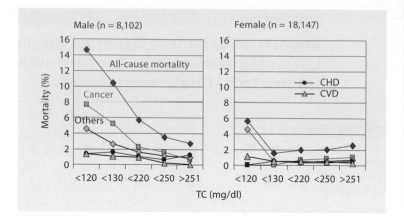

Hukui citizens (95% of the total were ≥40 years old), who had a general medical examination during 1986–1989 (n = 26,249), were followed for 5 years. The numbers of male subjects in each quintile of TC value were 157, 193, 6,099, 849 and 338 from the lowest TC group.

Conclusion: *Contrary to generally accepted understandings, high TC and obesity (data not shown here) were not the risk factors, but rather low TC and leanness were the risk factors for longevity in this population (translated).*

Our comments: These results indicated that high TC was a beneficial factor, not a risk factor, for various kinds of diseases and longevity in this general population. It should be noted that the highest TC group included all with a TC value ≥251 mg/dl. Although the mortality data were not adjusted for possible confounding factors, similar conclusions were derived from other studies (fig. 30, 32–35).

Fig. 32

General population in Ibaraki Prefecture was followed for 5.2 years after annual health checkups

Data taken from Irie et al. [2001].

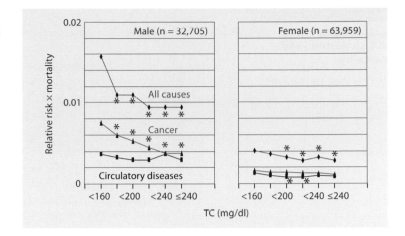

Participants of annual health checkups in 1993 (32,705 male and 63,959 female, aged 40–79 years) were followed for 5 years and 2 months. The Cox's proportional hazards model was used. There were 2,937 deaths (384 from stroke, 242 from CHD and 1,305 from cancer). The total number of deaths from CHD was relatively small, and multivariate relative risk values in the three TC groups of <180, 180–220 and >220 mg/dl were 1.0, 1.2 and 1.6(*) for men and 1.0, 1.1, 1.9(*) for women, respectively.

Conclusions: *Smoking, usual alcohol intake, hypertension, BMI, TC, HDL, blood glucose, carnitine and urinary protein were significantly associated with mortality. We obtained the new finding that serum creatinine level is a significant predictor of mortality from all cardiovascular diseases in Japanese men and women, and that the multivariate relative risk in female moderate alcohol drinkers vs. non-drinkers is significantly elevated for death from all causes (data not shown). The results of our study are useful for planning health care education and services.*

Our comments: The data presented in this report tell us more than the authors' conclusion. All-cause and cancer mortalities were significantly lower in the higher TC groups both in men and women, and the association between TC and all circulatory diseases was not significant. The CHD mortality was positively associated with TC but the number of deaths was relatively small for accurate analysis.

Fig. 33

**Ibaraki Study –
HDL cholesterol and
mortality**

Data taken from Irie et al. [2001].

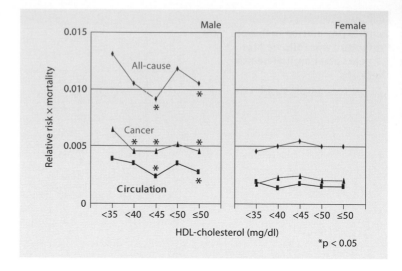

Conclusion: *Low HDL was a predictor for all-cause, cancer, all circulatory disease and CHD mortality for men.*

Our comments: Only the lowest HDL group (<35 mg/dl) exhibited higher relative risks for these diseases but the association was very small, if any, above 35 mg/dl in men. No significant association was observed between HDL and mortalities from these diseases in women.

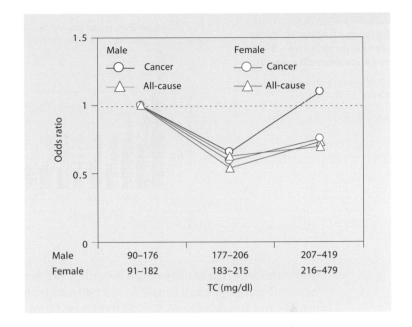

Fig. 34

Toda citizens followed for 10 years

Data taken from Shibata et al. [1995].

The baseline data were obtained during 1979–1981 (before the studies shown in fig. 30–35) for 1,202 male and 2,054 female subjects (general population). The mortality during the 10-year follow-up period was analyzed by multivariate logistic regression analysis.

Conclusion: *The present study revealed the inverse relationship of TC and HDL to all-cause and cancer mortalities in men.*

Our comments: Similar results from other populations are summarized in table 7.

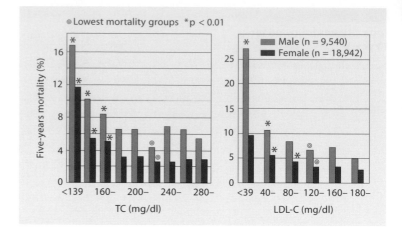

Fig. 35

Koriyama-Isesaki Study on general population – 5-year all-cause mortality and TC

Adapted from Ogushi [2004], kindly provided by Dr. Y. Ogushi.

Inhabitants who had health examinations in 1999 were followed up until 2004. The numbers of male and female were 9,540 and 18,942, respectively. The ages for male and female were 65.5 $\pm$ 10.0 (SD) and 62.6 $\pm$ 10.8 (SD). Each death was inspected with the Basic Resident Register in Japan. The subjects were divided into groups according to their results of TC or LDL-C in the baseline. Mortalities in 5 years were calculated for the groups. The group with the lowest mortality was specified for TC or LDL-C individually. The death rates of other groups were compared with the ones of the lowest groups. Fisher's exact test was used to calculate statistical probabilities.

Conclusion: *Mortalities were higher in the groups with TC values lower than 180 mg/dl, or LDL-C values lower than 80 mg/dl (male) and 120 mg/dl (female), compared with those of the lowest cholesterol group (p < 0.01). High TC and LDL-C were not risk factors for all-cause death among Japanese inhabitants.*

Our comments: The author's group proposes TC levels of 140–259 mg/dl (male) and 160–279 mg/dl (female), and LD-C levels of 72–180 (male) and 82–192 (female) mg/dl as safe ranges, which are much higher than the TC values of 140–219 mg/dl that are currently adopted by the majority of workers in Japanese medical fields. We interpret the data to indicate that only the TC levels below 179 mg/dl and LDL-C levels below 119 mg/dl are predictors of shorter survival; there is no need to set upper safe limits in TC and LDL-C values among general populations.

Almost the same results have been reported for general populations of Moriyama City [Tsuji et al., 2004].

3.4.

Interpretation of Reports on Korean, British and Japanese American Men

Fig. 36

The higher the TC, the lower were the cancer and all-cause mortalities at least up to 251 mg/dl – A large-scale follow-up study on Korean men

Reproduced from Song et al. [2000], with permission from Oxford University Press.

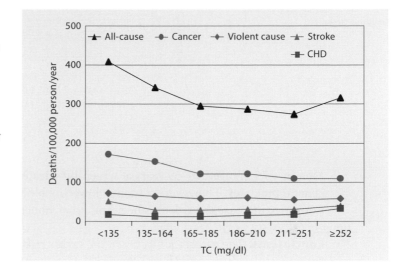

Subjects were Korean men (n = 482,472, aged 30–65 years) who had a baseline health examination in 1990. There were 7,894 deaths during the 6.4-year follow-up period. A low TC (<165 mg/dl) was associated with high all-cause mortality. The CHD mortality was high in the highest TC group (≥252 mg/dl). Mortalities from liver and colon cancer were significantly associated with a very low TC level (<135 mg/dl).

Conclusion: *The cholesterol level related with the lowest mortality ranged from 211 to 251 mg/dl, which was higher than the mean TC level of the subjects in the study.*

Our comments: The subjects were male public servants and teachers who had taken multi-phasic health examinations, not a high-risk group for CHD. In this group, cancer is the leading cause of death and CHD mortality is 5% of the total deaths. The profiles are very similar to those of Japanese general populations shown in figures 30–35.

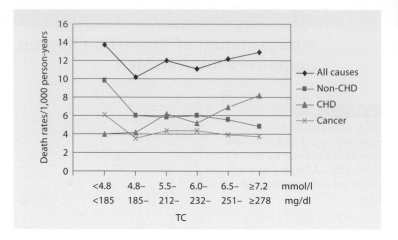

Fig. 37

Middle-aged British men from 24 towns followed for 14.8 years – British Regional Heart Study

Reproduced from Wannamethee et al. [1995], with permission from the BMJ Publishing Group.

Men (n = 7,735) aged 40–59 years at screening selected at random from the 24 general practices. There were fears that lowering blood cholesterol concentrations may increase the risk of cancer and other noncardiovascular deaths.

Conclusions: *The association between the excess risk of cancer and other non-cardiovascular disease and lower blood cholesterol concentrations (<4.8 mmol/l, 185 mg/dl) is produced by preclinical cancer, chronic ill health, smoking, and heavy drinking. In this series excess deaths in men with lower cholesterol concentrations occurred in the first 5 years of follow-up. There is no valid reason for changing current policies on lowering blood cholesterol concentrations.*

Our comments: In this population, the relationship between TC and all-cause mortalities reflects mainly those from CHD and cancer; CHD mortality was higher with higher TC levels, but cancer mortality was the highest in the lowest TC group. Two major factors made the interpretation of the results slightly complicated, i.e. the proportion of FH and the age group selected (40–59 years old). Because the age group over 60 years old was excluded from the subjects, the characteristics of aged populations, i.e. no significant positive association of high TC with CHD and inverse association of high TC with cancer mortality, are not reflected in this study. The ratio of CHD mortality at TC of ≥278 to that at <185 mg/dl is roughly 2, and this relatively small ratio is likely to reflect a relatively low proportion of FH in the highest TC group because the age group below 40 years was excluded; a high TC group under 40 years of age is likely to include greater proportions of FH than in older populations. Therefore, the proportion of FH in this population is critical for drawing the conclu-

　Prevention of Coronary Heart Disease

sion mentioned above. In general populations in Japan and Korea, CHD mortality is relatively low, and the relationship between TC and mortality from cancer or all causes was clearer than in this population (data shown above).

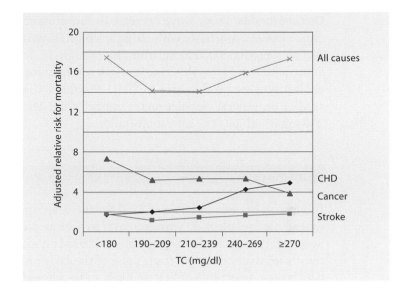

Fig. 38

Serum cholesterol and mortality among Japanese-American men. The Honolulu (Hawaii) Heart Program

Data taken from Stemmermann et al. [1991].

Hawaiian men of Japanese ancestry (45–68 years of age) followed up for 18 or more years after a baseline examination showed a quadratic distribution of death rates at different levels of TC. Mortality from cancer progressively decreased and mortality from CHD progressively increased with rising levels of TC. There was a positive association between baseline TC levels and deaths from CHD at 0–6, 7–12 and 13 years and longer after examination. The inverse relationship between cancer and TC levels was stronger in the first 6 years than in the next 6 years and, although still inverse, lost statistical significance after 13 years. Cancers of the colon and lung showed the strongest association with low baseline TC levels, while gastric or rectal cancer failed to show this association. Organ specificity and persistence of the inverse association beyond 6 years suggest that the nutritional demands of cancers may not entirely explain the inverse association with some cancers. The quadratic distribution of deaths in this cohort remained after CHD, stroke, and cancer were removed from the analysis (Multivariate RR adjusted for systolic blood pressure, cigarette smoking, alcohol consumption, BMI, and age at examination).

Conclusion: *For the entire period of observation, the lowest mortalities were found in men with TC levels between 180 and 239 mg/dl. Manipulation of TC levels below this level would not be desirable if this was to result in increased risk of death from cancer or other disease. This study does not rule out this possibility.*

Our comments: Progressive decrease in cancer mortality and progressive increase in CHD mortality with rising levels of TC were similar to those seen in British men (fig. 37). The relative risk of high TC for CHD was 2.9, and the contribution of FH to this risk is of interest because the aged population did exhibit progressive decrease in CHD mortality (fig. 25). As in many other studies on general populations, a range of TC levels from roughly 180 to 240 mg/dl exhibits the lowest mortality.

Summary

In Japan, a law was in effect in 1992 for the health promotion of aged people that advised people at 40 years and over to have medical examinations for free. Several follow-up studies were published based on such examinations, and the results were very different from those of the earlier hospital-based surveys, in which the proportions of FH are likely to be relatively high. The population-based surveys clearly indicated that CHD is not positively associated with TC and that high TC is a predictor for low cancer and all-cause mortalities.

Similarly, aged populations (>50 years old) in Western countries exhibited relatively little association between TC and CHD, while cancer and/or all-cause mortality tended to be inversely associated with TC (Austrian VHM&PP Study and Framingham Study, Studies on elderly people shown in table 8). It should be noted that high TC is not a risk factor for general populations over 40–50 years old during which atherogenesis is in progress, and that high TC is a predictor of low cancer mortality and longevity among these populations.

Chapter 4

Pleiotropic Effects of Statins in the Prevention of Coronary Heart Disease – Potential Side Effects

Statins inhibit HMG-CoA reductase and thereby slow the formation of isoprenoid intermediates in cholesterol synthesis. The first inhibitor that was a registered medicine in Japan was of microbial origin and was withdrawn from the market for unknown reasons. Many other kinds of statins have been shown in Western countries to be effective for the primary and secondary prevention of CHD, and are generally accepted in medical fields as 'evidence-based medicines'. However, the effectiveness of statins in decreasing CHD risk seems less related to a hypocholesterolemic activity than to suppressing other isoprenoid intermediates (farnesyl pyrophosphate, geranylgeranyl pyrophosphate) that are likely to be involved in the beneficial effects and harmful side effects of statins.

CHD mortality is much lower in Japan than in the Western industrialized countries, and unexpected consequences were revealed in the first large-scale intervention trial with simvastatin. A relatively polar statin (pravastatin) was shown to be effective for the prevention of CHD events in Japan as well (MEGA Study). The side effects of statins after chronic administrations may be more serious than presently estimated.

4.1.
Effectiveness of Statins for the Primary and Secondary Prevention of Coronary Heart Disease Proved in Western Countries

Table 10

Summary of intervention trials with statins which were performed in Western countries

Cholesterol level	Primary prevention		Secondary prevention		
	normal	hyper	normal	hyper	low HDL-C
Risk reduction, %	37	31	24	34	22
Drug	lovastatin	pravastatin	pravastatin	simvastatin	gemfibrozil
Trial	CAPS	WOS	CARE	4S	VA-HIT
	Downs et al. [1998]	Shephard et al. [1995]	Sacks et al. [1996]	Scandinavian Simvastatin Survival Study Group [1994]	Rubins et al. [1999]

Our comments: The beneficial effects of statins were proved and their side effects were reported to be tolerable during several years of interventions, although a couple of other statins were withdrawn from the market.

Fig. 39

Statins decrease CHD events regardless of the TC and LDL levels

Data taken from Kendall and Nuttall [2002].

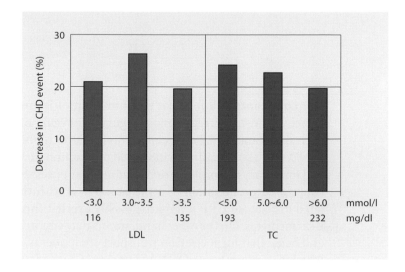

Conclusion: *Simvastatin produced an approximately 25% reduction in rate of major vascular events irrespective of previous or coexisting disease, irrespective of age and irrespective of gender. Of particular interest, simvastatin produced comparable reductions in vascular events irrespective of the prior LDL and TC.*

Our comments: The similar effectiveness in decreasing CHD events regardless of the TC and LDL levels suggests that the benefits obtained from statins are likely to be mediated by mechanisms other than lowering TC and LDL.

Fig. 40

Pleiotropic effects of statins exerted mainly through isoprenyl intermediates, and statins' side effects

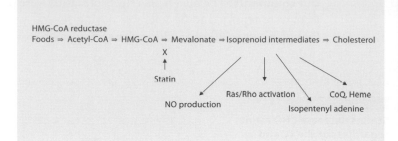

HMG-CoA reductase
Foods ⇒ Acetyl-CoA ⇒ HMG-CoA ⇒ Mevalonate ⇒ Isoprenoid intermediates ⇒ Cholesterol
X
↑
Statin
NO production   Ras/Rho activation   CoQ, Heme
Isopentenyl adenine

Our comments: Statins slow the formation of isoprenoid intermediates and thereby tend to decrease many regulatory events of isoprenylated proteins, such as Ras and Rho. Beneficial results of statin inhibition are: less suppression of NO synthesis (leading to increased production of vasodilating NO) and less inflammatory cell proliferation (leading to suppressed atherogenesis as well as suppressed proliferation of some cancers). Anti-inflammatory activities (related to reduced CRP level) are possibly exerted through lowered isoprenoid intermediates. On the other hand, statins may suppress the production of CoQ and heme that form electron transport components, and isopentenyl adenine which is a minor base of tRNA. The side effects of statins, such as rhabdomyolysis, carcinogenesis and impaired brain functions, may also be mediated through alterations of these intermediates. Although not noted for humans, decreased numbers of offspring and decreased growth rates of offspring have been noted in animal experiments with statins.

4.2.

Potential Side Effects of Statins

Table 11

Carcinogenicity of hypocholesterolemic drugs

Data taken from Newman and Hulley [1996].

Drugs	Animal	Relative exposure*	Neoplasia observed
Lovastatin	rat	2–7	hepatocellular carcinomas
	mouse	1–4	stomach papillomas, hepatocellular carcinomas, pulmonary adenomas
Pravastatin sodium	rat	6–10	hepatocellular carcinomas
	mouse	0.5–5	malignant lymphomas
Simvastatin	rat	29–45	liver and thyroid tumors
	mouse	15	liver carcinomas and adenomas, lung adenomas
Fluvastatin sodium	rat	25–35	thyroid adenomas and carcinomas, forestomach papillomas
	mouse	2–7	forestomach papillomas
Cholestyramine	rat	–	enhanced intestinal carcinogenesis
Clofibrate	rat	5	liver tumors
Gemfibrozil	rat	1.3	liver cancer

* Ratio of dose (area under the curve) in animal experiments to clinical dose.

Conclusions: *Extrapolation of this evidence of carcinogenesis from rodents to humans is an uncertain process. Longer-term clinical trials and careful post marketing surveillance during the next several decades are needed to determine whether cholesterol-lowering drugs cause cancer in humans. In the meantime, the results of experiments in animals and humans suggest that lipid-lowering drug treatment, especially with the fibrates and statins, should be avoided except in patients at high short-term risk of CHD.*

Our comments: Although carcinogenic effects of statins have been interpreted to be negative in a meta-analysis of the intervention trials performed in the Western countries (table 10), several years of follow-up may be too short to evaluate their carcinogenic effects accurately in clinical trials. Cancer mortality is not an independent measure but is affected by mortalities from other causes, and high CHD mortalities in these trials in the Western countries might have obscured potentially higher cancer mortality (tables 11, 12; fig. 41).

Table 12

Fatal heart attack and breast cancer rates in the CARE trial (Pravastatin)

Original data in Sacks et al. [1996].

	Statin group	Control group	Difference
Death from heart attack	96/2,081 (4.6%)	119/2,078 (5.7%)	–1.1%
Case of breast cancer	12/290 (4.5%)	1/286 (0.3%)	+4.2%

The data in the table above were taken from Ravnskov [2000], with permission from the author.

Conclusion by Sacks et al. [1996]: *These results demonstrate that the benefit of cholesterol-lowering therapy (risk reduction from 5.7 to 4.6%) extends to the majority of patients with coronary disease who have average cholesterol levels.*

Evaluation by Ravnskov [2000]: *The merits (a marginal reduction of death from coronary heart disease) and demerits (increased incidence of breast cancer) by statin treatment as shown in this table need to be critically evaluated.*

Our Comment: Statins suppress carcinogenesis when administered together with carcinogens in animal experiments. However, statins are carcinogenic by themselves. Statins' carcinogenic activity may be brought about by causing ischemia and inflammation through reduced productions of heme and CoQ, by interfering protein synthesis through reduced minor tRNA component (isopentenyl adenine), by suppressing immunological potency through inhibition of cholesterol synthesis required for proliferation of cytotoxic T cells, and/or by other mechanisms (chap. 9).

Fig. 41

Pravastatin in elderly individuals at risk of vascular disease (PROSPER): A randomized controlled trial

Data taken from Shepherd et al. [2002].

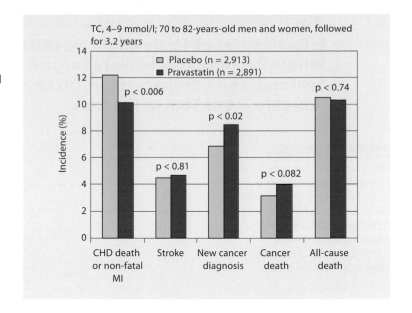

Men (n = 2,804) and women (n = 3,000) aged 70–82 years with a history of, or risk factors for, vascular disease were assigned to pravastatin (n = 2,891) or placebo (n = 2,913). Baseline TC ranged from 4 to 9 mmol/l (154–347 mg/dl). Follow-up was 3.2 years on average. Pravastatin lowered LDL-C by 34% and reduced the incidence of the primary endpoints, the hazard ratio being 0.81 (p = 0.014). CHD death and nonfatal MI was also reduced (0.81, p = 0.006). Stroke risk was unaffected (1.03, p = 0.8). New cancer diagnoses were more frequent on pravastatin than on placebo (1.25, p = 0.020). However, incorporation of this finding in a meta-analysis of all pravastatin and all statin trials showed no overall increase in risk.

Interpretation: *Pravastatin given for 3 years reduced the risk of coronary disease in elderly individuals. PROSPER therefore extends to elderly individuals the treatment strategy currently used in middle-aged people.*

Our comments: People in the elderly group are likely to have more precancerous cells in their body than young people, and carcinogenic activity of statins shown in animal experiments may have been revealed in this human study on elderly people. Obviously, longer-term follow-up is necessary to evaluate carcinogenic activity in clinical trials because of the presence of latent period in carcinogenesis.

4.3.

Effectiveness of Statin Might Be Different in Japan Where People Eat Relatively Large Amounts of Seafood, and Coronary Heart Disease Events Are Roughly 1/4 of That of the USA

Fig. 42

A large-scale cohort study of the relationship between TC and coronary events with low-dose simvastatin therapy in Japanese patients with hypercholesterolemia (>220 mg/dl) (J-LIT, Japan Lipid Intervention Trial)

Data taken from Matsuzaki et al. [2002].

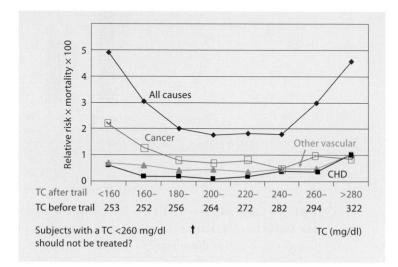

Men and women aged 35–70 years (n = 41,801) were followed for 6 years.

Conclusion: *A reasonable strategy to prevent coronary events in Japanese hypercholesterolemic patients without prior CHD under low-dose statin treatment might be regulating the serum lipid concentrations to at least <240 mg/dl for TC, <160 mg/dl for LDL, 300 mg/dl for TG (triacylglycerol), and >40 mg/dl for HDL (data not shown).*

Our comments: The original data were presented by setting the relative risk at a TC value of 256 mg/dl as 1 for each cause of death. In order to visualize the contribution of each cause to all-cause deaths, the relative risk (presented in the original data) was multiplied by the mortality at this TC value.

An unexpected U-shape was observed for all-cause and CHD mortalities. Cancer and all-cause mortalities were the highest in the lowest TC group. The higher cancer mortality with lower TC may be related to an apparent carcinogenic effect of statin (fig. 41). The carcinogenic effect of statin might have become evident in Japan where CHD mortality was relatively low.

It is often argued that higher cancer mortality observed in lower TC groups is the consequence of decreased TC associated with cachexia that was brought about by the development of tumors. However, this interpretation is unlikely because cancer death at the lowest TC group was only 0.4% of the subpopulation, which would not affect the average TC value of this subgroup appreciably.

Another factor to be considered in this trial is the presence of a very high proportion of FH in the subjects (2.5% compared to ~0.2% in average Japanese). The incidence of CHD of FH (heterozygote) is reported to be >10-fold higher (table 6). The number of FH was 1,059 and the number of CHD events was 355. Because FH is more resistant to chronic statin treatment, it is likely that most of FH are included in the high TC groups (>260 mg/dl) after the statin treatment. Subtracting the calculated contribution of FH from the groups (>260 mg/dl), we obtain a curve similar to those of general populations in Japan (fig. 30–35), that is, it is very likely that high TC is not positively associated with CHD in non-FH subjects.

Mortalities were higher in the groups with TC values below 180 mg/dl after the treatment but these groups had average TC values of 256 mg/dl before the treatment. Despite the conclusions deduced by the authors, patients with TC values of around 260 mg/dl should not be treated with the statin.

This J-LIT study had no control group, but an Area-Matched Control Study in Relation to J-LIT was reported separately (fig. 11). A clinical doctor pointed out that the mortality rate was 2.0%/5.39 years in the J-LIT intervention while it was only 1.4%/6 years in the control group, and that the J-LIT Study revealed no beneficial effects of simvastatin on this population. Unless these confounding factors are adjusted in the statistical analyses, the conclusions by the authors may require some revisions.

Table 13

Management of elevated cholesterol in the primary prevention group of adult Japanese (MEGA Study): results of the randomized MEGA study with pravastatin

Data taken from Nakamura [2005], with permission from Dr. H. Nakamura.

Group	Results of MEGA study with pravastatin in Japan (1,000 person years)		Hazard ratio (95% CI)	p value
	dietary advice (n = 3,966)	dietary advice and pravastatin (n = 3,866)		
CHD	101 (5.0)	66 (3.3)	0.67 (0.49–0.91)	0.01
MI	33 (1.6)	17 (0.9)	0.52 (0.29–0.94)	0.03
Fatal	3	2	–	–
Nonfatal	30	16	–	–
Stroke	652 (3.0)	50 (2.5)	0.83 (0.57–1.21)	0.33
Cerebral infarction	46	34	–	–
Intracranial	14	16	–	–
Bleeding unidentified	2	0	–	–
All-cause death	79 (3.8)	55 (2.7)	0.72 (0.51–1.01)	0.005
Cardiovascular	18	11	0.63 (0.30–1.33)	–
Noncardiovascular	61	44	0.74 (0.50–1.09)	–

A prospective, randomized, open-labeled, blinded endpoints study. Men (n = 3,966) and women (n = 3,866) with a TC value of 220–270 mg/dl, but without history of CHD, were randomized to 'dietary intervention' group and 'dietary intervention plus pravastatin' group, and CHD events were followed for ≥5 years in average. Risk reduction was 33% with 11% decrease in TC.

Conclusion: *The effectiveness of pravastatin for the prevention of CHD events was shown with a regular dose of 10–20 mg/day without overt side effects (translated).*

Our comments: The dietary advice is likely to be essentially the same as that used in the Area-Matched Control Study in Relation to Japan Lipid Intervention Trial (J-LIT) (fig. 11), in which the 'dietary advice' was a risk factor with the highest hazard ratio of 2.8. Although the dietary advice was made to both groups, compliance to the dietary advice in the statin group might be lower than in the control group because pravastatin effectively decreased TC values. The possible difference in the compliance to dietary advice might have affected the outcome in a favorable direction. The follow-up period (5 years) may not be enough for this age group (55–59.6 years old on average) to reveal the carcinogenic activity suggested from the PROSPER study (fig. 41, 42; table 12) and animal experiments (table 11). The hypocholesterolemic activity of pravastatin was relatively small (up to 11.5%) and was similarly effective to the lower (<240 mg/dl) and higher TC groups (≥240 mg/dl), the hazard ratio being 0.63 and 0.70, respectively. These results, together with those of other statins, indicate that pravastatin exerted its effect through isoprenyl intermediates and that the observed effectiveness of statins does not validate the interpretation that high TC is the major causative factor for CHD.

4.4.

Side Effects of Statins and Socio-Economic Aspects of Statin Treatment

The brain is the organ with the highest content of cholesterol, which is synthesized in the brain but not of plasma LDL origin. Lipophilic statins tend to exhibit hypocholesterolemic activities and suppress brain cholesterol synthesis more effectively than hydrophilic statins (e.g. pravastatin). Statins were once proposed to prevent Alzheimer disease but now the opposite effects are suspected. Hedgehog signaling involves covalent modification of protein with cholesterol, and plant alkaloids interfering this signaling induce brain anomaly and teratogenicity [Ingraham, 2001]. A case report also described cognitive disorder caused by statin-treatments [King et al., 2003].

As CHD events in Japan are roughly 1/4 of that in the USA, it is calculated that the life span of Japanese would be extended by 15 days if statin treatment was continued for 30 years for high TC subjects. On the other hand, cost-benefit aspects have been analyzed and its ineffectiveness has been discussed [Krut et al., 1998].

Summary

Although the effectiveness of statin treatment for the primary and secondary prevention of CHD has been well established, we are very much concerned about the consequences of chronic administration of statins. It takes many more years for the development of cancers in humans than the periods of clinical follow-up of statin treatments performed so far. Statin administration is not recommended for young women because of possible side effects on the next generation. The actions of statins have been revealed to be pleiotropic, but our current knowledge is not enough to predict all their side effects. For example, cholesterol content in the brain is much higher than in most other tissues and brain cholesterol is synthesized within the brain. Studies to evaluate the consequences of chronic inhibition of isoprenoids and cholesterol synthesis in the brain have just started in relation to Alzheimer disease, hedgehog signal transduction and other regulatory dysfunctions.

Because we already know safer, less costly and more effective means to prevent CHD events, we must carefully discuss the choice of interventions to be taken. The following chapters describe the need for using biomarkers that are linearly linked to fatal CHD events in future clinical epidemiological studies, and they emphasize the importance of changing currently adopted lipid nutrition advice to new directions that have a more rational basis.

Chapter 5

Objective Measures of the Pathology of Coronary Heart Disease

The first four chapters of this book illustrate how biochemical data on the serum lipids that are used as traditional 'risk factors' correlated poorly with observed CHD mortality. CHD death is a clear endpoint for epidemiologists to use in large-scale clinical studies, but it usually occurs only after decades of progressive inflammatory disease that eventually leads to thrombosis, ischemia and arrhythmia. Extensive pathobiological studies of Americans in the PDAY study (fig. 45, below) show clearly that inflammatory vascular injury begins in young people and accumulates steadily over decades before ending with a CHD death. As a result, long-term prevention of atherosclerosis should begin in childhood or adolescence. At an age when CHD deaths begin to be frequent, a very high percent of American people already have significant inflammatory vascular disease that is difficult to reverse. Because the rate of disease progression is much greater for people with American life-styles than for the Japanese life-style, it is important to explore the causes for this difference.

Although more expensive and difficult to obtain, objective measures of vascular pathology are more useful than mortality data in monitoring the progressive development of CHD, predicting CHD mortality and associating these events with preventive dietary and pharmaceutical interventions. To help design more effective prevention studies, this chapter examines current evidence on the usefulness of alternative endpoints as biomarkers, which might be linearly linked to fatal CHD events, and measures that are more effective than TC in clinical epidemiology. This chapter considers several alternatives to TC as possible biomarkers for monitoring risks of CHD in future clinical intervention studies.

In addition to blood biomarkers of inflammation (chap. 8), this chapter includes some physiological measures of cardiovascular and cerebrovascular integrity that are useful for monitoring preventive dietary interventions.

1   Postmortem examination, which is useful in defining the pathological subtypes of stroke.
2   Pulse wave velocity (PWV), which measures the speed of pulse wave through blood vessels; harder (atherosclerotic) vessels have faster velocity. Unfortunately, the number of research reports with sufficient PWV data is limited.
3   Angiograms, which show the degree of narrowing in blood vessel diameters due to arteriosclerosis.
4   Heart rate variability, which is measured using Holter monitor and reflects the status of autonomic control systems. It is used to evaluate the possible dietary interventions for the prevention of CHD.
5   Intima-media thickness (IMT), which is a measure of inflammatory proliferative atherosclerosis associated with CHD. A recent report on 5-lipoxy-

genase polymorphisms confirmed that eating dietary ω6 FAs is positively correlated with IMT in the subjects studied.

6 Hypertension, which results from several mechanisms for vascular integrity related to balance among dietary ω3 and ω6 PUFAs.

Only published data on studies with an appreciable number of subjects followed for relatively long periods were selected for presentation here, and more data of this type are needed.

5.1.

Hisayama Study (Japan) in Which the Cause of Death Was Defined by Postmortem Examination

Fig.43

Changes in TC levels in three examinations (1961, 1974 and 1988) in Hisayama town

Adapted from Fujishima [2001], with permission from Medical Review Co., Ltd.

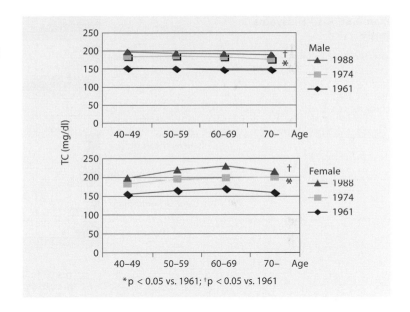

A series of studies on the population of Hisayama town (which is representative of average Japanese lifestyle) had high accuracy of defining the type of death, which was determined by postmortem examinations. The health of people in Hisayama town (Fukuoka Prefecture), ≥40 years old, was followed for 32 years, and the results were reported for four age groups. The average TC values increased progressively from the first (1961), to the second (1974) and third (1988) examinations (as it also did in the whole Japanese population).

Conclusion: *TC values tended to increase slightly with age in women (40–69 years old), but not much in men.*

Our comments: Although certain events are often discussed as the 'cause of death', the type of death event from circulatory diseases in this study was carefully defined for three major types, but there was little controlled evidence about the mechanisms causing each type of death. The small changes in TC over time may reflect gradual changes in food composition and life style over that time period.

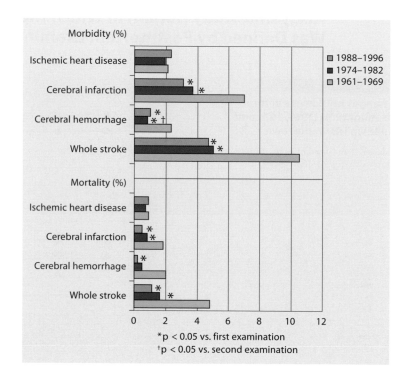

Fig. 44

Trends of mortality and morbidity from circulatory diseases in the three examinations – Hisayama Study

Data taken from Fujishima [2001].

Mortality was summed in four 8-year periods and expressed as deaths per 1,000 people per year. The number of subjects of the 1st, 2nd and 3rd populations was 1,621, 2,053 and 2,649, respectively. The CHD mortality in men was positively correlated with Atherogenic Index [(TC-HDL-C)/HDL-C] (data not shown in the original report). During this period (1961–1988), average TC values increased both in men and women, mortalities from cerebral infarction, cerebral bleeding and all stroke decreased significantly while that of CHD was unchanged.

Conclusion: *The incidence of hypertension, the most important risk factor of cardiovascular diseases, decreased steadily from the 1st to the 3rd examination, but the decreasing rate of stroke incidence has been diminishing recently, and the incidence of CHD remained unchanged during this period. It is highly possible that the preventive effect of hypertension management is compensated for by the increasing metabolic anomalies such as hyperlipidemia and glucose intolerance (a part of the original summary was translated).*

Our comments: Increase in TC values was not associated with increase in CHD death rates but was associated with decrease in incidence of stroke during the 30 years. Very low mortality from stroke in Okinawa prefecture, Japan (well known for their longevity) was suggested to be the major basis for their longevity, and it was associated with a high intake of pork after World War II (data not shown).

Table 14

Relative risk of TC for cerebral infarction and its subtypes in Hisayama – 32-year follow-up study

Data taken from Tanizaki et al. [2000].

Relative Risk [a]	Total cerebral infarction	Lacunar	Athero-thrombotic	Cardio embolic
Men, event	n = 144	n = 81	n = 29	n = 31
Age-adjusted RR	1.1	1.2	1.1	1.0
Multivariate RR	_[b]	_[b]	_[b]	_[b]
Women, event	n = 154	n = 86	n = 33	n = 25
Age-adjusted RR	1.1	1.2	1.4	0.6[c]
Multivariate RR	_[b]	_[b]	_[b]	0.8[c]

[a] Risk for an increase of 1 mmol/l TC; [b] TC was found not to be the risk factor; [c] $p < 0.05$.

This community-based, prospective cohort study in Hisayama, Japan, included men (n = 707) and women (n = 914) aged 40 years and over, who were followed for 32 years from their 1st examination in 1961 until 1994, and the cause of death was defined by postmortem examinations.

Conclusion: *TC showed an inverse correlation with mortality from cardioembolic infarction only in women (p < 0.05).*

Our comments: In Japan, the biomarker, TC, was not correlated with total cerebral infarction or its subtypes in men or in women, but TC correlated inversely with cardioembolic infarction. The 'Cholesterol Hypothesis' does not fit the deaths in Hisayama.

5.2.

Pathobiological Studies of Americans in the PDAY Study

Fig. 45

Association of CHD risk factors with the intermediate lesion of atherosclerosis in youth (PDAY Study)

Reproduced from McGill et al. [2000], with permission from Lippincott Williams & Wilkins.

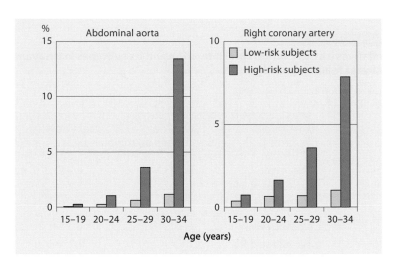

Subjects (n = 2,876) aged 15–34 years who died of external causes (accidents, homicides or suicides) within 72 h of injury were autopsied within 48 h of death. High-risk groups were characterized by non-HDL cholesterol of ≥160 mg/dl, HDL cholesterol of <35 mg/dl, smoking, hypertensive (≥110 mm Hg), and BMI of ≥30.

Conclusion: *Association of risk factors with raised fatty streaks became evident in subjects in their late teens, whereas associations of risk factors with raised lesions became evident in subjects aged >25 years. These results are consistent with the putative transitional role of raised fatty streaks and show that CHD risk factors accelerate atherogenesis in the second decade of life. Thus, long-range prevention of atherosclerosis should begin in childhood or adolescence.*

Our comments: This study clearly indicates that prevention of atherosclerosis should begin at younger ages. However, relative contributions of risk factors to raised lesions and their causal relationship need to be evaluated carefully. For example, hypertension analyzed by intimal thickness and an algorithm to estimate mean arterial pressure was characterized to be a risk factor in this study, but it may be the effect of atherosclerosis. Other studies revealed unfavorable results of antihypertensive drugs during interventions for CHD, e.g. MRFIT Study (fig. 9), J-LIT Area-Matched Control Study (fig. 11), and ALLHAT Study [The ALLHAT Officers and Coordinators for the ALLHAT Collaborative Research Group, 2002]. Another factor to be estimated is the proportion of FH; high LDL/HDL group in these young generations may include more than the average proportion of FH.

5.3.

Pulse Wave Velocity Correlated Positively with Ischemic Changes in Electrocardiogram but Not with Total Cholesterol Levels

Fig. 46

Positive correlation of PWV with ischemic ECG changes

Reproduced from Suzuki et al. [1987], with permission from the Japan Atherosclerosis Society.

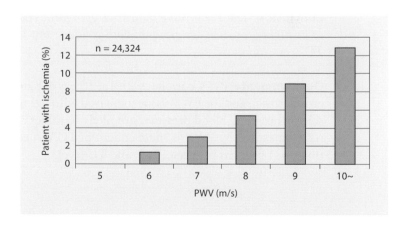

Conclusion: *The incidence of ischemic ECG changes increased exponentially with PWV values in men, suggesting that a progressive rise of PWV value preceded ischemic ECG change.*

Our comment: This measure of vascular dysfunction may be useful as an early intermediate biomarker prior to a severe clinical CHD event or death.

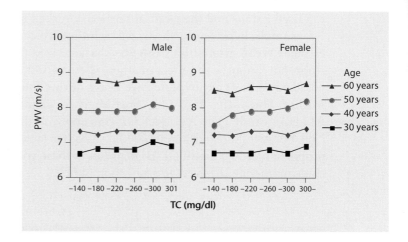

The incidence of ischemic ECG changes with each serum lipid value showed weak positive correlations for TC of women in their 30s to their 50s and stronger correlations for men.

Conclusion: *The PWV value indicative of organic changes in the aorta tended to be higher for each age group of both genders, but it was not correlated with TC.*

Our comments: PWV correlates with stiffness of blood vessels (degree of atherosclerosis), and the PWV value increased with aging. Factors other than vessel stiffness, e.g. blood pressure, are likely to affect the PWV, but atherosclerosis measured by PWV is independent of TC values; no significant association was observed between PWV and TC level.

5.4.

Angiograms Estimate Progression of Atherosclerosis That Was Independent of Total Cholesterol Values

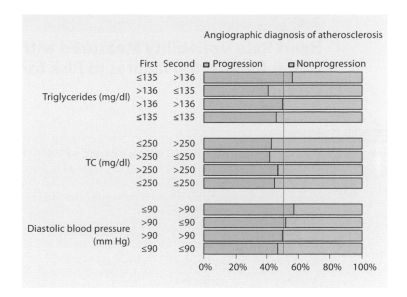

Fig. 48

Progression of coronary atherosclerosis during the first and second catheterization was not correlated significantly with the changes in TC values

Data taken from Kramer et al. [1983].

Subjects (n = 385) with unoperated coronary artery disease and two coronary angiograms separated by at least 1 month are evaluated. Progression/regression definitions: 26 arterial segments are compared at first and second catheterization, and progression is said to occur if any of the following arterial changes are noted at second study: (1) <100% to 100%; (2) <70% to 90% or more; (3) <30% to >50%, and (4) 20% or more increase of obstruction in any vessel narrowed by 50% or more at first study.

Conclusion: *Progression did not relate significantly to many traditional risk factors (including TC) at first and second catheterization.*

Our comments: Progression was associated with decreasing left-ventricular function (data not listed here). Conclusions from such objective measures have been disregarded in the Cholesterol Hypothesis. Biomarkers more closely linked to fatal CHD events need to be used in clinical epidemiology and in controlled interventions intended to prevent fatal coronary events. For example, the LACMART clinical trial used coronary angiography and intravascular ultrasound to monitor the regression of plaques in FH patients following lipoprotein apheresis that successfully removed toxic oxidized LDL, although the follow-up period was 1 year [Matsusaki et al., 2002].

5.5.

Heart Rate Variability Measured with 24-Hour Holter Monitoring Relates to Risk for Coronary Heart Disease

Table 15

Heart rate variability as a measure of CHD events – Eating ω3 PUFA (fish) relationship

Data adapted from Christensen et al. [2001].

ECG parameter	Modifiable factors			Nonmodifiable factors
	medication	ω3 PUFA related	life style	
RR Interval	β-blocker[a]	DHA[a], ω3 PUFA	tobacco[a]	age [a]
SDNN	β-blocker[b]	DHA	tobacco[a]	MI[a]
SDNNindex		EPA, DHA, ω3 PUFA	tobacco[a]	
SDANNindex	β-blocker[a]	DHA	tobacco[a]	
RMSSD	β-blocker[b]	EPA[a], ω3 PUFA[a]		age[b]
PNN50	β-blocker[b]	EPA, DHA, ω3 PUFA[a]		

[a] $p < 0.01$; [b] $p < 0.05$ in linear multiple regression analysis.
PUFA = Polyunsaturated FAs.

Linear multiple regression analysis used heart rate variability (HRV) indices as dependent factors with 291 subjects referred for coronary angiography after ischemic heart disease was suspected. The 24-hour heart rate variability was analyzed. RR = Average of normal RR interval; SDNN = standard deviation of normal RR intervals in the entire 24-hour recording; SDNN index = mean of SD of all normal RR intervals for all 5-min segments of the 24-hour recording; SDANN index = SD of the mean RR intervals measured in successive 5-min periods; pNN50 = percentage of successive RR-interval differences ≥50 ms; RMSSD = square root of the mean of the sum of the squares of differences between adjacent intervals.

Fish intake was negatively correlated with anomalies in ECG parameters measured with a 24-hour Holter monitor. In this population, those who ate fish also tended to drink more red wine, but no significant correlation was detected by multivariate analysis between the intake of red wine and ECG parameters.

Conclusions: *The close positive association between ω3 PUFAs and HRV in patients suspected of having ischemic heart disease may indicate a protective effect of ω3 PUFAs against sudden cardiac death (SCD). This may partly explain the reduction in SCD observed in humans with a modest intake of ω3 PUFA. Wine intake was also positively correlated with HRV, but this correlation was no longer significant after controlling for the cellular level of ω3 PUFA.*

Our comment: After multivariate analysis in this population, fish intake was associated with less risk for CHD as measured by heart rate variability, which reflects the status of autonomic control systems. Polyphenols in red wine are anti-oxidants and are presumed to prevent LDL oxidation, but their protective effects on CHD were not proven here.

5.6.
Intima-Media Thickness Relates to Dietary Fatty Acids and 5-Lipoxygenase Genotype

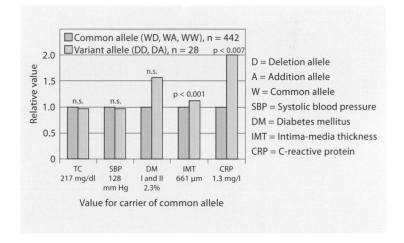

Fig. 49

Major CHD risk factors according to 5-lipoxygenase promoter genotype, and correlations between IMT and dietary FAs

Data adapted from Dwyer et al. [2004].

D = Deletion allele
A = Addition allele
W = Common allele
SBP = Systolic blood pressure
DM = Diabetes mellitus
IMT = Intima-media thickness
CRP = C-reactive protein

A variant genotype (lacking the common allele) was found in 6.0% of the cohort of 470 healthy, middle-aged women and men from the Los Angeles Atherosclerosis Study. IMT was 12% greater and the level of CRP (C-reactive protein, an inflammatory marker) was 2-fold higher in the carriers of variant alleles compared with that in carriers of common allele, although no significant difference was observed in traditional risk factors (TC, blood pressure and diabetes mellitus).

Conclusion: *People with the variant lipoxygenase alleles (but not those with the common allele) had IMT significantly associated with intake of both LA and AA. In contrast intake of ω3 marine fats was significantly inversely associated with IMT, and saturated and monounsaturated FAs had no significant association.*

Our Comment: The variant lipoxygenase alleles appear to increase the severity of inflammatory eicosanoid-mediated events, allowing a dietary imbalance of ω6 over ω3 fats to increase the severity and extent of IMT that develops from such inflammatory events.

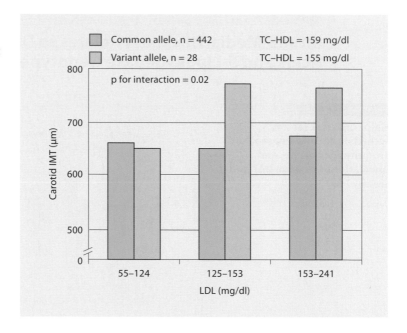

Fig. 50

Carotid IMT in the groups with common allele or variant allele according to LDL

A part of figure 3 in Dwyer et al. [2004], reproduced with permission from *New England Journal of Medicine.*

       Prevention of Coronary Heart Disease

Conclusion: *LDL levels did not differ significantly between the carriers of the two variant alleles and those of the common allele. However, the LDL level was a more potent atherogenic factor among carriers of two variant alleles than among carriers of the common allele.*

Our comment: The observed difference in IMT (~12%) was relatively small, and there was no significant difference in the LDL levels between the carriers of variant alleles and those of common allele. A possible mechanism for lipoxygenase action to support inflammatory events is by way of lipoxygenase-dependent metabolites regulating endothelial cell expression of PGHS-2 and its subsequent inflammatory events, a process that is down-regulated by the ω3 FA, EPA [Ait-Said et al., 2003]. Whether or not this mutation inactivates the 5-LPX is to be defined, and the number of variants (n = 28) needs to be increased to draw a definitive conclusion.

Fig. 51

Atherosclerosis and ω3 FAs in the populations of a fishing village and a farming village in Japan – PWV, IMT, plaque number and serum FAs

Data taken from Yamada et al. [2000].

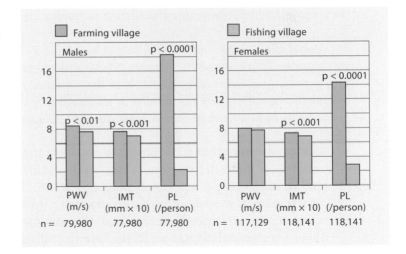

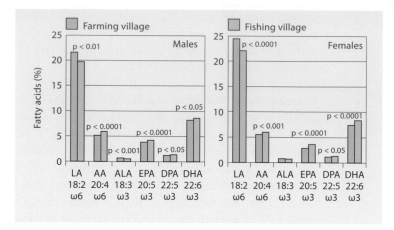

Fig. 52

Atherosclerosis and ω3 FAs in the populations of a fishing village and a farming village in Japan – PWV, IMT, plaque number and serum FAs

Data taken from Yamada et al. [2000].

The effect of dietary habits on atherosclerosis was investigated in a fishing village (n = 261) and a farming village (n = 209) in Mie Prefecture (central Japan). Life style, PWV of the aorta, IMT of the carotid artery, and atherosclerotic plaques (PL) were measured by ultrasonography.

Conclusion (major outcome): *There is a striking 5- to 8-fold difference in the number of atherosclerotic plaques between the populations in both men and women. Evaluation of the ω3 FAs over the combined populations reveals a negative association with the number of plaques in the common carotid while the ω6 FAs show a weak positive association with plaques.*

Our comments: Among the objective measures used in this paper, only the number of atherosclerotic plaques was strikingly different between the fishing and farming villages. The proportions of AA (ω6), EPA (ω3) and DHA (ω3) in serum lipids were significantly greater in the fishing village but the plaque number, the incidence of cerebral infarction and that of angina pectoris were more than severalfold greater in the farming village. The proportion of ω6 HUFA (%) in men was greater in the fishing village (29.4) than in the farming village (27.8) because of the presence of a small amount of AA in fish oil, which was often observed as pointed out in this paper. Thus, the causal relationship between the dietary FAs and atherosclerotic plaques (PL) is not explained simply in this study.

Prevention of Coronary Heart Disease

Hypertension – Relationships to Coronary Heart Disease and Stroke

Hypertension is positively correlated with CHD and stroke mortality, and is generally accepted as one of the major risk factors for eventual CHD mortality. The term 'risk factor' generally indicates an associated occurrence, but it does not automatically mean that reducing a risk factor will be beneficial, as it would be for a causal risk factor. Although hypertension has a logical causal relationship to hemorrhagic stroke, its relationship to thrombotic or ischemic events remains to be established. Both CHD and hypertension might independently be made worse by some processes initiated by some common dietary causes. By now, scientists can see that neither CHD nor hypertension is likely to be due to dietary cholesterol. However, both disorders may relate to an unbalanced dietary intake of ω6 over ω3 PUFAs that causes unbalanced signaling from ω6 over ω3 eicosanoids. A meta-analysis of 36 clinical trials involving 2,114 subjects showed that supplementing with high intakes of fish oil gave lower blood pressures, especially for older and hypertensive subjects [Geleijnse et al., 2002]. However, the observed antihypertensive effect of fish oil is marginal (1.5–1.7 mm Hg reduction in double-blind trials), hence fish oil is likely to exert preventive effects on CHD mainly through other mechanisms (chap. 6).

A relationship of dietary ω6 and ω3 fats with the prevalence of ST-T changes in ECG and medication for hypertension was evident among hundreds of Japanese in Brazil and Japan [Mizushima et al., 1997]. This study suggests a possible association between fish intake and reduced cardiovascular risk, through the beneficial effects of taurine and ω3 PUFA and a habitual low intake of calories and fat. Taurine is probably a surrogate marker of seafood intake because its reported cholesterol-lowering activity is significant only in huge amounts (2% of diet) in animal experiments, which is nearly impractical in human nutrition.

Summary

After World War II, Japanese intake of fats and oils increased gradually and so did blood TC values. In the typical town of Hisayama, CHD mortality was relatively unchanged, but the mortality from ischemic stroke actually decreased (fig. 43, 44). The PDAY study emphasizes the importance of starting prevention from younger ages, but relative contributions and causal relationships of the listed risk factors need to be carefully evaluated. Many objective measures of vascular function are related to the development of CHD and atherosclerosis (fig. 47, 48, 51), but they have no significant correlation with blood TC values. LDL was associated with atherogenesis among carriers of 5-lipoxygenase variant alleles, but not among carriers of common allele, although the number of the former was relatively small (n − 28). The association may be due to greater inflammatory oxidative conditions within carriers, which converts phospholipids in LDL to potent toxic vasoactive agents. Obviously, we need more data on objective measures with appreciable numbers of subjects followed for relatively long periods.

ω3 Fatty Acids Effectively Prevent Coronary Heart Disease and Other Late-Onset Diseases – The Excessive Linoleic Acid Syndrome

Epidemiological studies on Greenland natives (Inuit) and Danes, gave clues to biochemical mechanisms by which fish oil ω3 FAs prevent atherosclerosis-related diseases [Lands, 1986]. Intervention trials and large-scale and long-term follow-up studies led to the definitive conclusion that CHD and its related symptoms can be prevented effectively by supplementing diets with fish oil.

6.1.
Three Different Types of Highly Unsaturated Fatty Acid

Table 16

Three types of highly unsaturated FA (HUFA) are formed by competing metabolic steps

Saturated and monounsaturated FAs form the ω9 type of HUFA
Saturated FA (S) ⇒ monounsaturated FA (M) ⇒ ⇒ ⇒ (⇒ Mead acid, 20:3ω9)
Animal fats   animal fats
Palm oil   high-oleic vegetable oil (olive oil, canola)
(made from protein, carbohydrate)

Linoleic acid forms the ω6 type of HUFA
Linoleic acid ⇒ γ-linolenic acid ⇒ dihomo-γ-linolenic acid ⇒ arachidonic acid ⇒ (22:5ω6)
LA (18:2ω6)   GLA (18:3 ω6)   DGLA (20:3ω6)   AA (20:4ω6)
Grain   (small amounts in meats)
Vegetable oils   ↓   ↓
ω6 eicosanoids (inflammatory mediators)

α-Linolenic acid forms the ω3 type of HUFA:
α-Linolenic acid ⇒ ⇒ eicosapentaenoic acid ⇒ ⇒ docosahexaenoic acid
ALA (18:3ω3)   EPA (20:5ω3)   DHA (22:6ω3)
Leaves   seafood   seafood
Seafood   phytoplankton   phytoplankton
Phytoplankton   ↓
Perilla oil, Linseed oil   ω3 eicosanoids   (brain and retina)

Our comments: S and M are synthesized in our body but LA and ALA must be eaten (essential FAs).

Ingested FAs compete with each other as they are desaturated and elongated in our body to form three different types: ω3, ω6 and ω9. However, no inter-conversion occurs among the three different types.

Various foods contain different proportions of these FAs, and tissue ω6 and ω3 FA compositions change depending on the choice of foods, because these FAs can only come from foods.

Various physiologically active materials (eicosanoids, inflammatory mediators) are formed from LA by way of AA and DGLA (LA cascade, or AA cascade), and from ALA by way of EPA and DHA.

Metabolism of ω6 and ω3 FAs leading to eicosanoids acting through their respective receptors is competitive. Therefore, not only the absolute amounts of ω6 and ω3 FAs but their relative proportions are important for the diet-tissue-disease relationship.

High proportions of ω6 to ω3 FAs in tissue HUFA is a consequence of choosing foods with high ω6/ω3 ratios, and the high proportions correlate with high inflammatory, thrombotic and arrhythmia tendencies, a major risk for CHD deaths, cancer proliferation, allergic hyper-reactivity and other inflammatory diseases.

Fig. 53

Relationship among dietary FAs, membrane phospholipids and lipid mediators (eicosanoids), and their competitive aspects

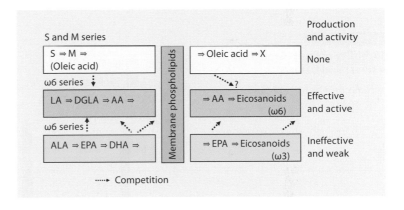

Arachidonic acid (AA) is converted to thrombotic ω6 thromboxane A_2 (TXA$_2$) in platelets and anti-thrombotic ω6 prostaglandin I_2 (PGI$_2$) in vascular endothelial cells. When the proportion of AA increases in membrane phospholipids, the TXA$_2$/PGI$_2$ ratio is elevated. EPA is converted to eicosanoids by cyclooxygenases less effectively than AA, and the EPA-derived eicosanoids (TXA$_3$ and leukotriene B$_{5)}$) are biologically less active than the corresponding products from AA. Another important aspect is that the FAs of the three different types (table 16) compete with each other at many enzymatic steps and at eicosanoid receptors.

Conclusion: *There is no strong rationale to recommend increasing both the ω3 FAs (fish oil) and ω6 FAs (high-LA vegetable oils) at the same time to lower CHD risk. Prolonged eating of high-LA vegetable oils is not hypocholesterolemic in long-term interventions run (fig. 5), and increasing their intake may support an over-production of ω6 eicosanoids that increase thrombotic, arrhythmic and inflammatory events.*

Our Comment: This chapter provides clinical evidence of benefits from eating ω3 fats. Pleiotropic actions of EPA and ω3 fats will be discussed more in chapter 8.

6.2.

Clinical Evidence that ω3 Fatty Acids Are Effective for the Primary and Secondary Prevention of Coronary Heart Disease

Fig. 54

Morbidity of Inuit (Greenland Natives) and Danes

Data taken from Kroman and Green [1980] and Lands [1986].

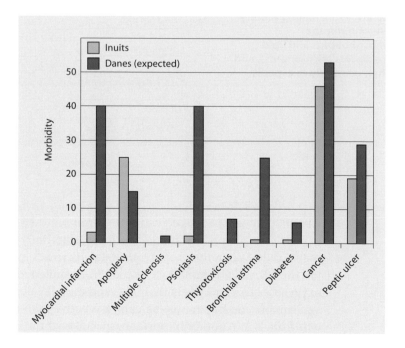

Prevention of Coronary Heart Disease

The study population (approximately 1,800 inhabitants) is one of the remaining whaling and sealing populations in Greenland. They were observed over the 25-year period 1950–1974 concerning the incidence of the diseases, which was based on all cases diagnosed in hospital during this period.

The disease pattern of the Greenlanders differs from that of West-European populations, having a higher frequency of apoplexy and epilepsy but a lower frequency or absence of acute myocardial infarction and others shown in this figure. The distribution of cancer types differs from that of the Danish population, but the total incidence of cancer is of the same magnitude.

Conclusions: *Further comparable studies should be performed in Greenlandic districts that are characterized by more profound changes in life style, in order to elucidate the effect of these changes on the disease pattern.*

Our comments: As compared with Danes, Inuit ate 2-fold more cholesterol than Danes, and the intake of S and M was over 30 en% in both populations. TC values were not much different but the degree of atherosclerosis was much less in Inuit, and the age-adjusted morbidity from CHD was below 1/10 that of Danes. The major risk factor for atherosclerosis-CHD is not cholesterol and animal fats, but high $\omega6/\omega3$ ratio of the diet (fig. 53; and chap. 8, fig. 75, 82).

Although the incidence of cerebral bleeding was significantly higher among Inuit, an increased bleeding tendency has not been reported in clinical studies with relatively large amounts of fish oil supplements. Instead, an impaired synthesis of collagen and elastin due to vitamin C deficiency has been postulated for the higher bleeding tendency in Inuit. The morbidity from cancer was not different between the two populations. However, the incidence of cancers common in the USA (but not smoking-related cancers) was very low in Inuit compared with Danes.

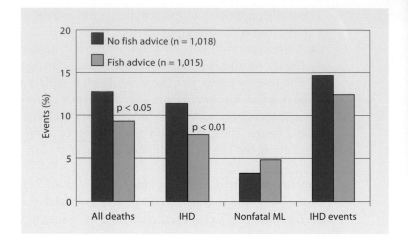

Fig. 55

Diet and Reinfarction Trial (DART)

Data taken from Burr et al. [1989].

Men under 70 years old (n = 2,033) who survived a myocardial infarction (MI) were allocated to receive advice on one of three dietary factors, followed for 2 years, and outcomes compared.

1 A reduction in fat intake and an increase in the ratio of P/S.
2 An increase in fatty fish intake.
3 An increase in cereal fiber intake.
   Results corresponding to 1–3 described above were:
1 No difference in mortality. TC reduction was 3–4%.
2 A 29% reduction in all-cause mortality.
3 A slightly higher mortality (n.s.).

Conclusion: *A modest intake of fatty fish may reduce mortality in men who have recovered from MI.*

Our comments: For the secondary prevention of myocardial infarction, dietary advice to reduce fat intake and raise P/S ratio of FAs has been essentially ineffective, but the effectiveness of eating fish was clearly demonstrated.

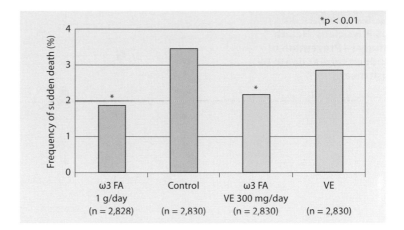

Fig. 56

GISSI-Prevenzion Investigators – Study for the treatment of myocardial infarction

Data taken from GISSI-Prevenzion Investigators [1999].

Patients surviving a recent (<3 months) myocardial infarction were randomly assigned supplements of ω3 PUFA (in this case, EPA and DHA) (1 g/day, n = 2,836), vitamin E (300 mg/day, n = 2,830), both (n = 2,830) or none (control, n = 2,828) for 3.5 years. Treatment with ω3 PUFA, but not vitamin E, significantly lowered the risk of all-cause death and cardiovascular death.

Conclusion: *Dietary supplementation with ω3 PUFA led to a clinically important and statistically significant benefit. Vitamin E had no benefit.*

Our comments: In the so-called free radical theory of atherogenesis and aging, lipid peroxides are presumed to decompose and increase free radicals resulting in increased oxidized LDL, hence anti-oxidants are protective for these diseases. Although EPA and DHA are very susceptible to auto-oxidation to form lipid peroxides in the air, these PUFAs were demonstrated to suppress CHD and aging in vivo.

Fig. 57

US Physicians' Health Study I – Prevention of sudden cardiac death by fish meal

Data taken from Albert et al. [1998].

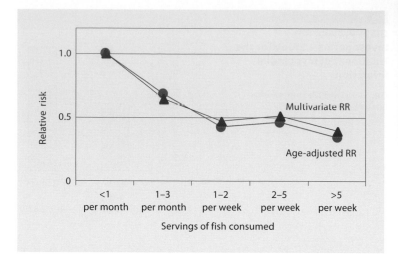

This study enrolled a total of 20,551 US male physicians 40–84 years of age and free of MI, cerebrovascular disease, and cancer. At baseline subjects completed an abbreviated, semiquantitative food frequency questionnaire on fish consumption and were randomized for aspirin and β-carotene treatment and then followed up to 11 years, controlling for age and coronary risk factors. Dietary fish intake was associated with a reduced risk of sudden death, with an apparent threshold effect at a consumption level of 1 fish meal per week (p for trend = 0.03).

Conclusion: *These prospective data suggest that consumption of fish at least once per week may reduce the risk of sudden cardiac death in men.*

Fig. 58

US Physicians' Health Study II

Data taken from Albert et al. [2002].

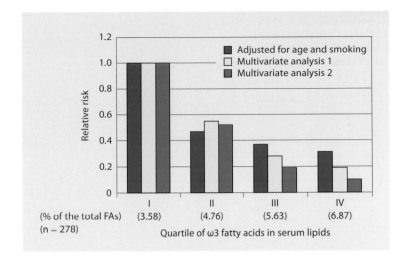

Apparently healthy men were followed for 17 years in the Physicians' Health Study (a prospective, nested case-control analysis). The FA composition of previously collected blood was analyzed by gas-liquid chromatography for 94 men in whom sudden death occurred as the first manifestation of cardiovascular disease and for 184 controls matched with them for age and smoking status.

Conclusion: *The serum levels of ω3 FAs obtained from fish are strongly associated with a reduced risk of sudden death among men without evidence of prior cardiovascular disease.*

Our comments: Large-scale prospective studies in US Physicians' Health Study I and II revealed that even a small amount of fish oil intake may be effective for the prevention of sudden cardiac death in the USA. Higher amounts of fish intake and reduced amounts of competitive ω6 FA (LA) are probably necessary to reduce the CHD events to the rate seen in Japan.

Fig. 59

US Nurses' Health Study I – Intake of fish and ω3 FAs and risk of stroke in women

Data taken from Iso et al. [2001]. Ref. Iso et al. [2002].

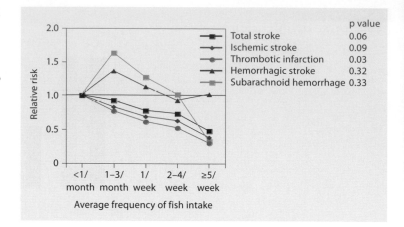

This longitudinal Health Study enrolled women aged 34–59 years in 1980 who were free from prior-diagnosed cardiovascular disease, cancer, and history of diabetes and hypercholesterolemia. At baseline, subjects completed a food frequency questionnaire including information on consumption of fish and other frequently eaten foods. The 79,839 women who met the eligibility criteria were followed up for 14 years.

Conclusion: *Higher consumption of fish and ω3 PUFAs is associated with a reduced risk of thrombotic infarction, primarily among women who do not take aspirin regularly, but it is not related to risk of hemorrhagic stroke.*

Our comments: This work clearly demonstrated the effectiveness of ω3 FAs for the prevention of thrombotic diseases. Another important conclusion derived from this work was that relatively large amounts of ω3 FA intake do not accelerate cerebral bleeding. The latter is particularly important because a possible association of high ω3 FA intake and apoplexy was suspected from epidemiological studies on Inuit and Danes (fig. 54).

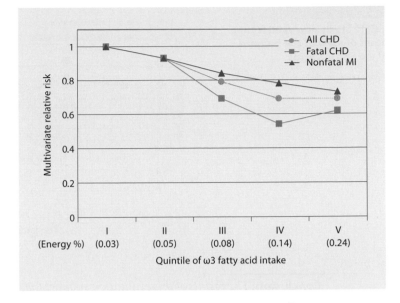

Fig. 60

**US Nurses' Health Study II –
Fish and ω3 FA intake and risk
of coronary heart disease in
women**

Data taken from Hu et al. [2002].

Conclusion: *Among women, higher consumption of fish and ω3 FAs is associated with a lower risk of CHD events, and CHD death.*

Our comments: Although intake of ω3 FAs was associated with significantly decreased CHD events, even the highest quintile of ω3 FA intake for this population of USA women was about half the average intake for Japanese, for whom CHD mortality was 1/4–1/5 the level of the USA.

Fig. 61

MRFIT Study analyzed from a point of view of the balance of ω6 and ω3 acids

Data taken from Dolecek et al. [1991].

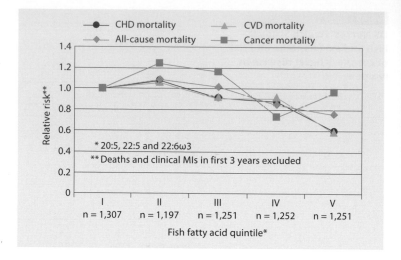

Conclusion: *The results of this evaluation support the hypothesis that FAs found primarily in fish oils protect against cardiovascular disease. They also suggest that the composition and balance of PUFA in the diet may influence mortality from cardiovascular disease and possibly various forms of cancer.*

Our comments: Fish oil ω3 FAs and α-linolenic acid (ALA) in the diet were inversely correlated with all-cause mortality, and the dietary ALA/LA ratio was inversely correlated with cancer mortality.

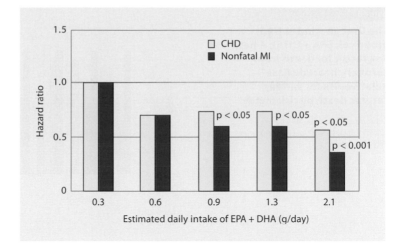

Fig. 62

Higher intake of ω3 FAs was associated with substantially reduced risk of CHD – The Japan Public Health Center-Based (JPHC) Study Cohort I

Data taken from Iso et al. [2006].

A total of 41,578 Japanese men and women aged 40–59 years who were free of prior diagnosis of cardiovascular disease and cancer and who completed a food frequency questionnaire were followed up from 1990–1992 to 2001. After 477,325 person-years of follow-up (11.5 years in average), 258 incident cases of coronary heart disease (198 definite and 23 probable myocardial infarctions and 37 sudden cardiac deaths) were documented, comprising 196 nonfatal and 62 fatal coronary events. The multivariable hazard ratios (HRs) and 95% confidence intervals in the highest (8 times per week, or median intake = 180 g/day) versus lowest (once a week, or median intake = 23 g/day) quintiles of fish intake were 0.63 (0.38–1.04) for total coronary heart disease, 0.44 (0.24–0.81) for definite myocardial infarction, and 1.14 (0.36–3.63) for sudden cardiac death. The reduced risk was primarily observed for nonfatal coronary events (IIR = 0.43 [0.23–0.81]) but not for fatal coronary events (HR = 1.08 [0.42–2.76]). Strong inverse associations existed between dietary intake of ω3 fatty acids and risk of definite myocardial infarction (HR = 0.35 [0.18–0.66]) and nonfatal coronary events (HR = 0.33 [0.17–0.63]).

Conclusion: *Compared with a modest fish intake of once a week and/or 20 g/day, a higher intake was associated with substantially reduced risk of coronary heart disease, primarily nonfatal cardiac events, among middle-aged persons.*

Our comments: The data suggest that increasing the intake of fish in the USA to the level ingested by Japanese would help reducing the incidence of CHD to the level of Japanese (1/4–1/5 that of the USA); the fish intake of the highest quintile in the USA is roughly half the level of average Japanese (fig. 60).

Fig. 63

The Omega-3 Index (red blood cell EPA + DHA): A new risk factor for death from coronary heart disease? – Relative risk for sudden cardiac death by risk factor

Reproduced from Harris and Von Schacky [2004], with permission from Elsevier.

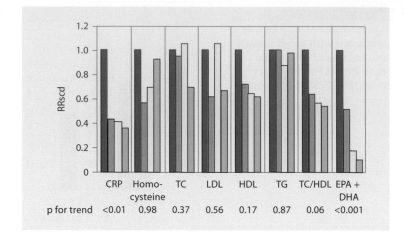

The relationship between the Omega-3 Index (blood level of EPA + DHA) and risk for CHD death, especially sudden cardiac death (SCD), was evaluated in several published primary and secondary prevention studies. The Omega-3 Index was adjusted for age; smoking status; assignment to aspirin/beta-carotene/placebo; BMI, history of diabetes, hypertension, or hypercholesterolemia; alcohol consumption; exercise frequency; parental history of MI before age 60 years; and *trans-* and monounsaturated FA intake. Results: The Omega-3 Index was inversely associated with risk for CHD mortality. An Omega-3 Index of $\geq 8\%$ was associated with the greatest cardio protection, whereas an index of $\leq 4\%$ was associated with the least.

Conclusion: *The Omega-3 Index may represent a novel, physiologically relevant, easily modified, independent, and graded risk factor for death from CHD that could have significant clinical utility.*

Our comments: As has been shown so far, red blood cell EPA + DHA was inversely associated with sudden cardiac death but neither of the cholesterol-related, traditional blood-borne parameters (TC, LDL, HDL, TC/HDL, TG) was associated with death. CRP (C-reactive protein) exhibited an inverse association with death.

Prevention of Coronary Heart Disease

Fig. 64

Effects of purified EPA (ethylester) on CHD events in Japanese patients with hypercholesterolemia: The Japan EPA Lipid Intervention Study (JELIS)

Reproduced from http://www.mochida. co.jp/dis/jelis/, with permission from the author (Dr. M. Yokoyama, University of Kobe, Graduate School of Medical Sciences) [Yokoyama et al., 2003].

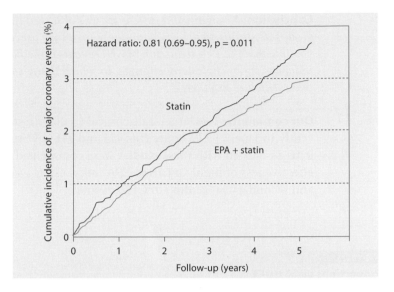

This study is a prospective, randomized, open-label, blinded end point trial including both primary and secondary prevention strata, with a maximum follow-up of 5 years. Its main purpose is to examine the clinical effectiveness of highly purified EPA (1,800 mg/day) given as an additional treatment to patients taking HMG-CoA reductase inhibitors for hypercholesterolemia. A primary end point is major coronary events: sudden cardiac death, fatal and nonfatal myocardial infarction, unstable angina pectoris including hospitalization for documented ischemic episodes, and events of angioplasty/stenting or coronary artery bypass grafting. Secondary end points include all-cause mortality, stroke, peripheral artery disease, and cancer. Baseline study composition comprises 15,000 participants (4,204 men and 10,796 women) in the primary prevention stratum and 3,645 (1,656 men and 1,989 women) in the secondary stratum. The minimum age is 40 years for men, women are required to be postmenopausal, and all patients must be ≤75 years of age. The mean age of participants is 61 years, and 69% are female. The schedule for plasma FA composition measurement is as follows: at baseline, at 6 months, and yearly thereafter. The mean baseline total and low-density lipoprotein cholesterol levels were 275 mg/dl (7.1 mmol/l) and 180 mg/dl (4.6 mmol/l).

Major outcome: Major coronary events were 3.5% in the control (statin) and 2.8% in the EPA group (statin + EPA), and the difference was 19%. In the secondary prevention study, a significant reduction of the major coronary events was observed. In both groups, LDL decreased by 26% while HDL remained unchanged.

Conclusion: *No significant difference was observed in the TC values of the two groups as statins were used commonly (primary prevention study). It is highly possible that the 19% reduction of CHD events is mediated through mechanisms independent of cholesterol changes, in which there is a rationale for taking the two medicines together.*

Our comments: Two large-scale clinical trials with statins were performed recently in Japan: J-LIT study (fig. 42) and MEGA Study (table 13). The problems associated with these studies were commented above. In this study, the effectiveness of highly purified EPA (ethyl ester) was shown for the primary and secondary prevention of CHD events.

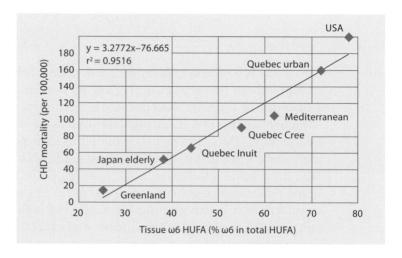

Fig. 65

Proportions of ω6 HUFA in tissue HUFA are highly positively correlated with CHD mortalities in different populations

Data adapted from Lands [2003].

Conclusion: *CHD mortality in different populations is strongly correlated (r > 0.98) with the proportion of ω6 HUFA in tissue HUFA. The observed proportions of ω6 in the HUFA ranged from 25 to 78% whereas the CHD mortality rates ranged from about 20 to 200 deaths per 100,000 populations.*

Our comments: CHD involves excessive ω6 eicosanoid actions in chronic and acute inflammatory processes in vascular walls that predispose people to fatal heart attacks as well as in the thrombosis and arrhythmia of the acute event. High proportions of ω6 in tissue HUFA will give high proportions in the non-esterified acids released during a stimulus. This, in turn, will give high rates of formation and action of ω6 eicosanoids, whereas low proportions will give low rates of formation. The balance of ω3 and ω6 acids in the diet influences

Prevention of Coronary Heart Disease

the balance of ω3 and ω6 HUFA in tissues and therefore the eventual balance of ω3 and ω6 eicosanoid actions in self-healing processes. Because CHD is a major cause of death in the USA, many drug treatments are marketed vigorously to meet the need to treat people and reduce an imminent risk. The wide diversity in values for this figure supports the idea of designing new dietary interventions to decrease the proportion of ω6 HUFA in tissue HUFA and to reduce CHD mortality.

Data taken from Nakamura et al. [2003].

Table 17

Serum FA levels, dietary style and coronary heart disease in three neighboring areas in Japan: The Kumihama study

Area	Kumihama Study			p value
	mercantile (n = 2,069)	farming (n = 7,571)	fishing (n = 2,944)	
Positive stress EKG*	290	238	34	<0.0001
Incidence of AMI*	24	30	8	0.45
Incidence of AP*	109	66	8	0.01
Smoking, %	12	17	28	0.01
LDL-C/HDL-C**	2.1	1.9	2.0	
LA, mg/dl	858	823	690	<0.0001
Stearic acid, mg/dl	217	246	198	<0.0001
EPA, mg/dl	90	84	97	0.07
AA/EPA	1.73	1.70	1.35	
% ω6 in total HUFA	38.3	38.4	35.0	#1
3 × (% ω6HUFA) – 75	40	40	30	

* Per 100,000; ** average/average; #1, calculated from the original data.

The factors influencing CHD mortality were investigated in a rural coastal district of Japan, comprising mercantile, farming, and fishing areas with distinct dietary habits. The incidence of CHD from 1994 to 1998, as well as coronary risk factors and serum FA concentrations, was prospectively examined. The incidence of angina pectoris was significantly (p = 0.01) lower in the fishing area than in the mercantile and farming areas. Blood pressure, physical activity, prevalence of diabetes, serum levels of uric acid and HDL-C were similar between the three areas. Total- and LDL-C levels were significantly lower but the smoking rate was markedly higher in the fishing area than in the other two areas. Serum levels of saturated FAs and ω6 PUFA were lowest in the fishing area, but ω3 PUFA did not differ significantly. The ω6:ω3 PUFA ratio was lowest and EPA:AA was highest in the fishing area.

Conclusion: *Although many previous studies have emphasized the beneficial effect of ω3 PUFA in preventing CHD, the present study indicated that a lower intake of ω6 PUFA and saturated FAs has an additional preventive effect on CHD even when the serum level of ω3 PUFA is high because of high dietary fish consumption.*

Our comments: This study showed greater incidence of undesirable acute MI, angina pectoris and post-stress ECG events linked with much higher circulating total fat, saturated FA and the ω6 LA that had no corresponding higher amount of balancing ω3 FAs.

Fig. 66

Differences in cardiovascular disease risk factors between Japanese in Japan and Japanese-Americans in Hawaii: The INTERLIPID study

Data taken from Ueshima et al. [2003].

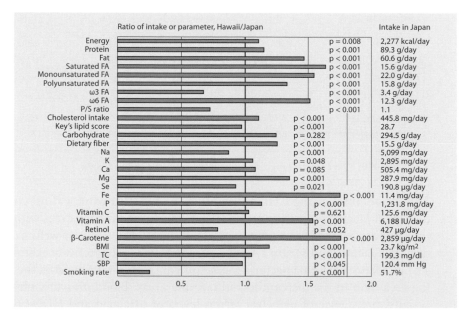

Prevention of Coronary Heart Disease

Despite higher Keys' score, smoking rate, and frequency of adverse blood pressure levels in Japan, coronary heart disease (CHD) incidence and mortality apparently remain substantially lower at all ages in Japan than in the US and other Western societies. CHD biomedical risk factors and dietary variables were compared in Japanese living in Japan and 3rd and 4th generation Japanese emigrants living a primarily Western lifestyle in Hawaii. Men and women aged 40–59 years were examined by common standardized methods – four samples in Japan (574 men, 571 women) and a Japanese-American sample in Hawaii (136 men, 131 women).

Outcomes: Average systolic (SBP) and diastolic (DBP) blood pressures were significantly higher in men in Japan than in Hawaii; there were no significant differences in women. The treatment rate of hypertension was much lower in Japan than Hawaii. Smoking prevalence was higher, markedly so for men, in Japan than Hawaii. Body mass index, serum total and low-density lipoprotein cholesterol, HbA1c, and fibrinogen were significantly lower in Japan than in Hawaii; high-density lipoprotein cholesterol was higher in Japan. Total fat and saturated FA intake were lower in Japan than in Hawaii. Polyunsaturated/saturated FA ratio and ω3 FA intake were higher in Japan than in Hawaii.

Conclusion: *Levels of several, especially lipid, CHD risk factors were generally lower in Japanese in Japan than in Japanese in Hawaii. These differences were smaller for women than men between Japan and Hawaii. They may partly explain lower CHD incidence and mortality in Japan than in Western industrialized countries.*

Our comments: These results do not support the interpretation that smoking, hypertension, low P/S ratio and high Keys' lipid score are major risk factors for CHD, but are consistent with the interpretation that low intake of ω3 and high intake of ω6 FAs are major risk factors. Although the observed differences in some parameters (Keys' lipid score, TC and SBP) are highly significant statistically, these differences appear to be too small to correlate with the large difference in CHD mortalities of the two countries.

6.3.

Recommendations to Decrease the Intake of Linoleic Acid (ω6) and Increase That of α-Linolenic Acid (ω3) and Oleic Acid (ω9) Were More Effective than Statin-Treatment for the Secondary Prevention of Coronary Heart Disease

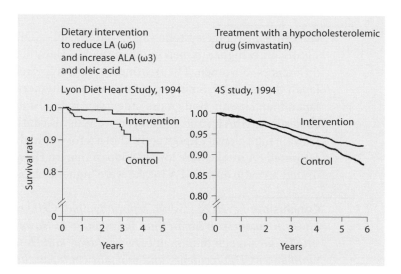

The effect of a Mediterranean ALA-rich diet was compared with that of a usual postinfarct prudent diet in a prospective, randomized single-blinded secondary prevention trial. After a first myocardial infarction, patients were randomly assigned to the experimental (n = 302) or control group (n = 303), and were followed for 4 years (Lyon Diet Heart Study).

Conclusions: *An ALA-rich Mediterranean diet seems to be more efficient than diets presently used in the secondary prevention of coronary events and death.*

Our comments: In the secondary prevention of CHD events, the Lyon Diet Heart Study achieved 70% risk reduction, whereas a similar trial with simvastatin reduced the risk only by 30%. Needless to say, a precise comparison is not possible because the selection of subjects and other experimental conditions were different. A recent overall meta-analysis of 11 studies of 13 cohorts including 222,364 individuals with an average of 11.8 years follow-up concluded that fish consumption is inversely associated with fatal CHD [He et al., 2004].

Summary

Effectiveness of fish oil ω3 FAs and vegetable oil ALA (ω3) for the primary and secondary prevention of CHD has been established. Because of competition between ω6 and ω3 FAs, reducing the intake of ω6 FAs (mostly LA in our current food environment) while increasing that of ω3 FAs is effective as shown in the Lyon Diet Heart Study and others (fig. 9–11; tables 4, 5). Some data inconsistent with this simple conclusion will be discussed in chapter 7.

Chapter 7

Why Isn't the Causal Relationship between Linoleic Acid and Mortalities from Coronary Heart Disease and Stroke Revealed by Clinical Studies?

Dietary linoleic acid (LA) is converted to AA, which is stored in tissue lipids and then converted to ω6 eicosanoids (LA cascade). An enhanced LA cascade may be a major cause of atherosclerosis (chap. 8) and cancer (chap. 9). A causal relationship between the LA cascade and cancer has been established in animal experiments. However, current limited epidemiological evidence has not necessarily supported a direct causal relationship for humans. Moreover, recent reports conclude that high serum LA is 'protective' (i.e. associated with lower risk) for stroke ([Iso et al., 2002]; see below) and breast cancer [Rissanen et al., 2003]. In these studies, stored serum samples were analyzed after storage for 1–9 years and 15 years, respectively. In the latter case, preferential breakdown of highly unsaturated FAs during storage can be suggested because the proportions of saturated FAs were 20–60% higher, LA was 10% lower and long chain PUFAs were 50–60% lower than in another Finland cohort during this period.

Tissue ω6 and ω3 FAs are of diet origin, and ingested fish oil increases EPA and DHA in serum lipids. However, intake of LA and serum LA were poorly correlated even when trained dietitians carefully evaluated their own ingested foods [Kuriki et al., 2003]. Similarly, intake of a large amount of ALA gives only a relatively small increase of ALA in serum lipids. This chapter will show that several metabolic factors lead to a very complex relationship between the intake and the serum level of 18-carbon PUFA. Serum LA is not a good measure of LA intake, just as TC is not a good measure of cholesterol intake.

First, a report claiming an inverse correlation between serum LA and stroke was examined carefully, and other reports on stroke-dietary fat relationship were examined.

Linoleic Acid and Stroke – Epidemiological Studies

Table 18

Linoleic acid, other FAs, and the risk of stroke – Baseline data

Data taken from Iso et al. [2002].

	n	Age	Male %	SBP	DBP	Hyper-tensive	BMI	Drinking g/day	Smoking %	TB mM	TG mM	Glc intoler	Diabetes %
Stroke cases were 197 in 6–9 years, and matched controls were selected													
Case	197	65.4	53	143[a]	83[a]	58[a]	24.0[b]	14.4	36[c]	5.9	1.42[c]	19	10[c]
Control	591	65.4	53	135	79	42	23.1	12.5	28	5.02	1.28	17	5
Serum fatty acid, % of total fatty acids													
	14:0	16:0	18:0	16:1	18:1	18:2	18:3	20:3	20:4	18:3	20:5	22:5	22:6
Nested serum samples from subjects without defining fasting conditions													
Case	1.2	23.8[b]	7.9	3.9[a]	21.4[b]	26.5	0.4	0.9	4.6	1.0	3.6	0.5	4.3
Control	1.1	23.0	7.9	3.6	20.9	28.2[a]	0.2	0.8	4.9[b]	1.1	3.5	0.5	4.3

[a] $p < 0.001$; [b] $p < 0.01$; [c] $p < 0.05$ vs. controls.

A cardiovascular risk survey was conducted during 1984–1992 for 7,450 participants aged 40–85 years (3 communities in Japan), and serum samples were stored at –80°C for 1–9 years. Three controls were selected for each of 197 stroke cases, matching for sex, age, community, year of serum storage, and fasting status. Compared with controls, stroke cases (total n = 197, hemorrhagic n = 75, and ischemic n = 122) had slightly (but significantly) higher blood pressure, BMI and serum TG.

Conclusions: *Compared with controls, stroke cases had in their total serum lipids similar proportions of ω3 FAs, lower proportions of LA and AA, and higher proportions of saturated and monounsaturated acids.*

Our comments: Although careful efforts were made in selecting controls matched to the cases, the baseline data show significantly higher known risk factors (such as blood pressure, hypertensive subjects, body mass index, drinking, smoking, TG level and diabetes) in the cases compared to controls. Because this study analyzed total serum FAs, the different LA contents in TG, phospholipids (PL) and cholesterol esters of the blood will greatly influence

Why Isn't the Causal Relationship between Linoleic Acid
and Mortalities from CHD and Stroke Revealed by Clinical Studies?
107

the average amount of LA reported. The slightly higher proportion of TG or lower proportion of PL and cholesterol ester (CE) in the cases compared to controls might have contributed to the slight difference of serum FAs observed between the cases and controls. The large number of people that were analyzed allowed very small differences observed to become significant from a statistical viewpoint, although the magnitude of the difference has no known physiological significance for any individual person.

Table 19

Linoleic acid, other FAs, and the risk of stroke – Correlation of stroke and serum FAs

Data taken from Iso et al. [2002].

	Quartile of serum fatty acids				
	I (low)	II	III	IV (high)	p value
Saturated acids, %	27.5	30.5	32.9	36.7	
OR[a]	1.0	1.28	1.28	1.84	0.02
OR[b]	1.0	0.96	0.82	1.07	0.45
Linoleic acid, %	22.4	26.4	29.9	33.7	
OR[a]	1.0	0.97	0.56	0.41	0.002
OR[b]	1.0	0.97	0.58	0.43	0.05

[a] Adjusted for BMI, smoking, drinking, hypertension, TC, TG, glucose index, gender, age, area, period of serum storage, fasting conditions.
[b] Additionally adjusted for monounsaturated and saturated or linoleic acid.

Conclusions: *A higher intake of LA may protect against ischemic stroke, possibly through potential mechanisms of decreased blood pressure, reduced platelet aggregation, and enhanced deformability of erythrocyte cells.*

Our comments: The odds ratio of the highest LA quartile was 0.43 and the p value (0.05) for the trend was not small enough for this size of study. Cholesterol ester is relatively enriched with LA, and higher baseline LA in the control may represent, at least in part, an elevated serum level of triglycerides and cholesterol esters. A high cholesterol level is associated with less risk for stroke in Japan (fig. 31, 44). A possible preferential breakdown of ω3 FAs during storage was corrected statistically. Many other factors affecting the serum levels of LA will be discussed below (fig. 69–74). Another aspect to be considered is the

Prevention of Coronary Heart Disease

presence of unidentified factors in vegetable oils that accelerate the onset of stroke in animals (chap. 8.4.). Serum levels of LA and ALA may be surrogate markers of such factors in part.

Despite the conclusion from this report, we do not advise to increase the intake of LA for the prevention of stroke until after the results of MRFIT Study (fig. 9), Helsinki Businessmen Study (fig. 10; table 4), Australian Secondary Prevention Study (table 15), Lyon Diet Heart Study (fig. 67) and Kumihama Study (table 17) are consistently and rationally interpreted.

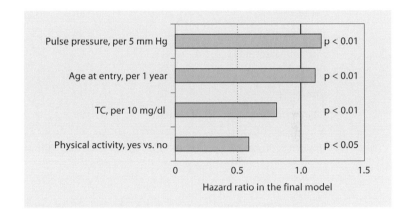

Fig. 68

Risk factors of stroke: A 40-year follow-up of the Corfu (Greece) cohort from the Seven-Countries Study

Data taken from Panagiotakos et al. [2003].

Corfu cohort (CVD-free men, n = 529, 49.7 ± 5.7 years old) were followed for 40 years from 1961. Death from hemorrhagic or thrombotic stroke was at age 74 years.

Conclusion: *Age, pulse pressure levels, physical activity (protective) and TC levels (protective) were significantly related with 40-year stroke mortality.*

Our comments: Besides the well-known risk factors (age and hypertension), high TC was revealed to be a protective factor for hemorrhagic and thrombotic stroke as it is recognized in Japan (fig. 44).

Table 20

**Dietary fat intake and risk of stroke in male US healthcare professionals:
14 years' prospective cohort study**

Data taken from Health Care Professional Study [He et al., 2003], with permission from the BMJ Publishing Group.

Dietary factor	Ischemic stroke		Hemorrhagic stroke	
	relative risk	p value	relative risk	p value
Total fat	0.91	0.77	1.16	0.83
Animal fat	1.20	0.47	0.90	0.90
Vegetable Fat	1.07	0.66	0.87	0.80
Saturated FA	1.16	0.59	0.99	0.85
Monounsaturated	0.91	0.83	0.95	0.82
Polyunsaturated FA	0.88	0.25	0.86	0.75
trans-FA	0.87	0.42	0.87	0.20
Dietary cholesterol	1.02	0.99	1.04	0.61

Health professional men (n = 43,732, 40–75 years old) followed up for 14 years.

Health professional men (n = 43,732, 40–75 years old) were followed for 14 years. Total stroke cases (n = 725), ischemic (n = 455), hemorrhagic (n = 125) and unknown (n = 145) were observed. After adjustment for confounders, no evidence was found that the amount or type of dietary fat affects the risk of developing ischemic or hemorrhagic stroke.

Conclusions: *The results do not support associations between intake of total fat, cholesterol or specific types of fat and risk of stroke in men.*

Our comments: There remains a possibility to be examined that the small range of ω6/ω3 ratios in ingested foods for this group eating 10–20 g PUFA/ day made tissue responses independent of dietary LA, and the possible beneficial effects of much lower ω6/ω3 ratios were not revealed in this study (fig. 69– 74).

7.2.
Factors Affecting the Variability of Serum Polyunsaturated Fatty Acid Compositions

Selective and competitive incorporation of unsaturated FAs into lipids – From animal experiments

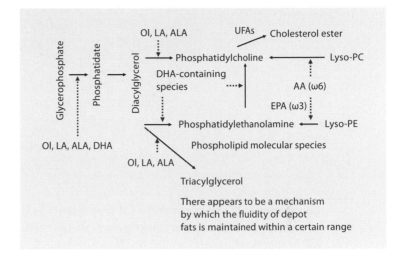

Our explanation: Various PUFAs ($\leq$18 carbon chains) are converted to HUFA ($\geq$20 carbon chains). These FAs are incorporated into lipids selectively and competitively, and the competitive effectiveness of FAs differ among different phospholipid classes and in different tissues.

Why Isn't the Causal Relationship between Linoleic Acid and Mortalities from CHD and Stroke Revealed by Clinical Studies?

111

Fig. 70

Relationship of ω6 eicosanoid precursors in tissue HUFA with its dietary precursor LA and competing ω3 HUFA in diet

Data adapted from Lands et al. [1992].

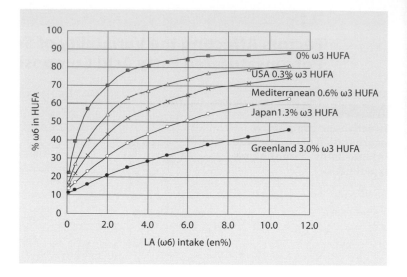

Conclusions: *We can predict the LA proportions in human serum TG (and adipose tissue), which depend linearly on its percent of dietary food energy, but the LA proportions in total serum lipids (which include TG, cholesterol esters and phospholipids) are not linearly dependent on dietary LA. An empirical hyperbolic equation fits extensive diet-tissue relationships in animals and humans published by earlier researchers, and it predicts for humans the likely proportion of ω6 HUFA (e.g. AA, dihomo-γ-linolenic acid) in plasma phospholipid HUFA that can respond to dietary LA intakes below 2 en% and not above 4 en%. It does not predict the proportions of 18-carbon PUFAs.*

Our comments: Failure to reveal a causal relationship between dietary LA and CHD in the USA could be explained by the competitive hyperbolic metabolism that converts LA into the ω6 HUFA, AA, which mediates important cardiovascular events. Many studies with experimental animals and hundreds of humans show that this metabolic conversion to tissue HUFA has a clear dose-response to dietary LA only below 1% of food energy, above which the accumulated tissue ω6 HUFA is relatively independent of widely varied LA intakes. Epidemiological studies usually involve people with voluntary LA intakes far above 1% of food energy and fail to observe a response related to dose. The proportion of ω6 HUFA in membrane phospholipids represents an intermediate stage of LA metabolism that leads to ω6 eicosanoid formation and action. The percent of ω6 HUFA in tissue HUFA is a function of the abundance of its precursor, LA, in the diet relative to the ω3 FAs (ALA, EPA and DHA), which are effective competitive inhibitors of the LA cascade. The proportion of ω6

Prevention of Coronary Heart Disease

HUFA in phospholipid HUFA is decreased by decreased dietary LA. However, this decrease is relatively ineffective when competing ω3 HUFA are absent or present at small amounts in the diet, as with the average US population. When LA is decreased from 8 en% to 4 en%, ω6 HUFA in phospholipids are expected to decrease significantly but they remain at a level above 70% as in the average USA population. However, with significant amounts of dietary EPA and DHA, as in the average Japanese, the proportion of ω6 HUFA could be decreased to a level below 50% by decreasing LA intake.

Correlations between the LA intake and the incidence of mammary tumors or CHD events are not widely shown in the epidemiological studies performed in the Western industrialized countries. This is likely when epidemiologists tend to make linear analyses of non-linear events. Also, many studies involve subjects who have only a narrow range for the proportion of ω6 HUFA in their phospholipid HUFA, even though a wide range is possible and actually occurs worldwide. The proportion of ω6 HUFA is a marker of LA and AA metabolism that predicts inflammatory, proliferative and thrombotic events. For example, most of the Americans in the MRFIT study had HUFA proportions near 75% ω6 in HUFA, and only one-fifth was near 60% ω6. Even though the lowest quintile had much lower CHD mortality, its effect on overall case-control comparisons for the whole study group was slight (but highly significant!). This means that the intake of LA should be decreased and that of ω3 HUFA should be increased to decrease the proportion of ω6 HUFA in phospholipids toward levels near 40% ω6 in HUFA, as usually seen with Japanese adults. Such a change in HUFA proportions is likely to lower significantly the incidence of CHD and other inflammatory diseases. Replacing high-LA vegetable oils with low-LA, high-ALA vegetable oils such as perilla seed oil and flaxseed (linseed) oil would also help decrease the proportion of ω6 HUFA in tissue HUFA. Because of the nonlinear response of tissue HUFA to dietary LA, one cannot conclude that LA intake is not associated with CHD.

Fig. 71

Optimizing DHA levels in piglets by lowering the LA to ALA ratio

Adapted from Blank et al. [2002]. This figure was kindly provided by the corresponding author (R. Gibson).

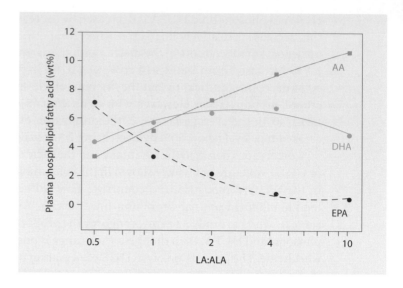

Effect of altering the LA to ALA ratio in the dietary fats was evaluated in 3-day-old piglets fed formula for 3 weeks. The LA/ALA ratios of the experimental formulas were 0.5:1, 1:1, 2:1, 4:1 and 10:1. The level of LA was held constant at 13% of total fats while the level of ALA varied from 1.3% (10:1 group) to 26.8% (0.5:1 group). Incorporation of the ω3 long chain PUFA (EPA and 22:5ω3) into erythrocytes, plasma, liver, and brain tissues was linearly related to dietary ALA. Conversely, incorporation of DHA into all tissues was related to dietary ALA in a curvilinear manner, with the maximum incorporation of DHA appearing to be between the LA-ALA ratios of 4:1 and 2:1. Feeding LA-ALA ratios of 10:1 and 0.5:1 resulted in lower and similar proportions of DHA in tissues despite the very different levels of dietary ALA (1.3 vs. 26.8% of total fats, respectively).

Conclusion: *These results are relevant to term infant studies in that they confirm our earlier findings of the positive effect on DHA status by lowering the LA-ALA ratio from 10:1 to 3:1 or 4:1, and they predict that ratios of LA-ALA below 4:1 would have little further beneficial effect on DHA status.*

Our comments: Increasing ALA at a constant level of LA resulted in roughly linear decrease in AA and increase in EPA, eicosanoid precursors. Among experimental animals, pigs are relatively similar to humans in lipid metabolism.

Effect of dietary ALA and its ratio to LA on platelet and plasma FAs and thrombogenesis in normolipidemic men

Adapted from Chan et al. [1993].

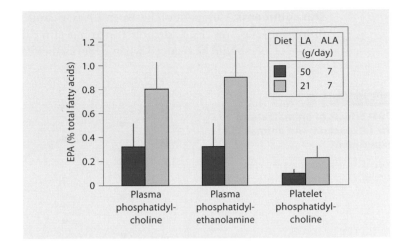

The effect of dietary ALA and its ratio to LA on platelet and plasma phospholipid (PL) FA patterns and prostanoid production were studied in normolipidemic men. The study consisted of two 42-day phases. Each was divided into a 6-day pre-experimental period, during which a mixed fat diet was fed, and two 18-day experimental periods, during which a mixture of sunflower and olive oil [low ALA content, high LA/ALA ratio (LO-HI diet)], soybean oil (intermediate ALA content, intermediate LA/ALA ratio), canola oil (intermediate ALA content, low LA/ALA ratio) and a mixture of sunflower, olive and flax oil [high ALA content, low LA/ALA ratio (HI-LO diet)] provided 77% of the fat (26% of the energy) in the diet. The ALA content and the LA/ALA ratio of the experimental diets were: 0.8%, 27.4; 6.5%, 6.9; 6.6%, 3.0, and 13.4%, 2.7, respectively. There were appreciable differences in the FA composition of platelet and plasma phospholipids. Nevertheless, oleic acid, LA and ALA levels in PL reflected the FA composition of the diets, although very little ALA was incorporated into PL. Both the level of ALA in the diet and the LA/ALA ratio were important in influencing the levels of longer chain ω3 FA, especially EPA, in platelet and plasma PL (the data are shown in this figure). Production of 6-keto-PGF1 alpha was significantly ($p < 0.05$) higher following the HI-LO diet than the LO-HI diet although dietary fat source had no effect on bleeding time or thromboxane B_2 production.

Conclusion: *Both the level of ALA in the diet and its ratio to LA were important in influencing long-chain ω3 FA levels in platelet and plasma PL and that prostanoid production coincided with the diet-induced differences in phospholipid FA patterns.*

Our comments: Competition between LA (ω6) and ALA (ω3), established in animal studies, was clearly demonstrated in a clinical trial, indicating that the intake of LA should be reduced when beneficial effects of EPA are expected.

Fig. 73

Dual effects of animal fats on the LA cascade – An animal experiment

Data taken from Broughton and Wade [2002].

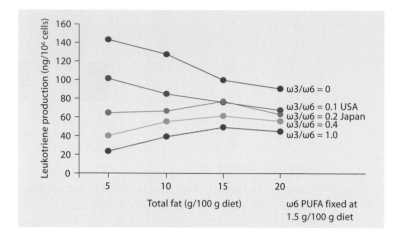

Note that the ω3/ω6 ratio is shown here while the opposite ratio is used in other data.

Dietary total fat at a fixed amount of LA as well as ω3/ω6 fat ratios influenced the leukotriene (C4, E4) production in mouse resident peritoneal cells. At a ω3/ω6 ratio of 0.1, the value comparable to that of average Americans (USA), increasing the amount of total fat (essentially animal fats) resulted in decreased productions of leukotrienes. However, the leukotriene production was increased along with increasing the amount of animal fats when the ω3/ω6 ratio was above 0.2, a value comparable to that of average Japanese.

Conclusion: *Although this study emphasizes that there are changes induced by varying both the total fat and ω3 PUFA contents of the diet, further studies could provide information on which to base general dietary recommendations. Future studies should further investigate the beneficial effect of various PUFA with regard to both eicosanoid-modulated physiologic functions and for optimizing plasma lipids.*

Our comments: We often put emphasis on the competition between ω3 and ω6 FAs, but this paper revealed a possible complex involvement of saturated and monounsaturated FAs in regulating the LA cascade and inflammatory tone. Simply decreasing the intake of animal fats without concomitant reduction of ω6 FAs might result in enhanced LA cascade.

Prevention of Coronary Heart Disease

Fig. 74

**Factors affecting the serum
LA level**

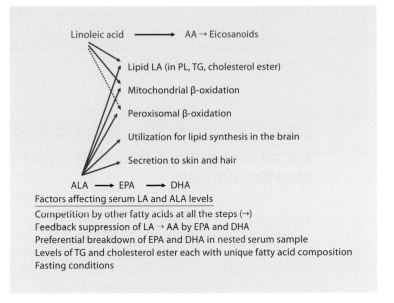

Factors affecting serum LA and ALA levels
Competition by other fatty acids at all the steps (→)
Feedback suppression of LA → AA by EPA and DHA
Preferential breakdown of EPA and DHA in nested serum sample
Levels of TG and cholesterol ester each with unique fatty acid composition
Fasting conditions

Our explanation: LA and ALA are not synthesized de novo in our body, and the LA and ALA levels in serum lipids do not reflect the ingested amounts of these FAs (even when trained dietitians carefully evaluated their own FA intake) [Kuriki et al., 2003]. Factors affecting serum LA and ALA levels may be listed as follows.

1 Serum LA derives from both food and depot fats; up to 15% LA in depot fat is common, depending on average habitual choices of foods. Usually (but not always), blood samples in fasting conditions are analyzed to avoid mixed results in epidemiological studies.
2 Competition among PUFAs occurs at many enzymatic steps involved in TG, cholesterol ester and phospholipid metabolism.
3 Competitions among PUFA occur during desaturation and elongation of LA and ALA, and the reactions are feedback suppressed by dietary and tissue long-chain HUFA (AA, EPA, DHA).
4 Each type of serum lipid (phospholipid, TG and cholesterol ester) has a 'characteristic' average FA composition that depends on the 'customary' average diet, and the serum levels of each type can differ with physiologic states. Although individuals can differ widely in food choices, clinical studies usually report average results from people with average diets customary of the nearby locality.

5   PUFAs with ≤18 carbon chains are effectively β-oxidized in mitochondria, but ALA is a better substrate than LA for the peroxisomal β-oxidation system. These systems are regulated with hormones related to energy metabolism and thermogenesis.

7.3.

Factors Other than Triacylglycerol in Vegetable Oils May Be Involved

Vegetable oils consist of mainly TG, but contain minor components such as phytosterols, lipophilic vitamins and many other minor components. Partially hydrogenated oils contain TG with *trans*-FAs and hydrogenated minor components. These minor components might have affected the observed dietary FAs-disease relationship, and some data that led to these considerations will be presented in chapter 8.4.

Summary

So far, some epidemiological studies have failed to reveal LA as a major risk factor for CHD and cancer, but this does not mean that intake of large amounts of LA is healthy. Our body is regulated in a complex manner and available information indicates a non-linear flow of dietary ω6 and ω3 FAs to sites on eicosanoid receptors. PUFAs with 18 carbons are metabolized significantly differently from the long-chain PUFAs or HUFAs. LA can be stored in depot fats up to 15% of the total FAs (in an order of >1 kg/person), and it takes a long time to change the ω6/ω3 balance in bodies of humans who are taking up to 20 g of LA daily.

Experimental animal experiments show that general metabolic selectivities of FA metabolism are similar for mice and rats, making it possible to examine details and conditions in which detrimental effects of excessive intake of LA are well established when evidence from human studies is not available. Even though the causal relationship between LA intake and disease has not been revealed in epidemiological studies performed in the USA, 'recommending LA intake up to 10 en%' (ATP III) should be avoided for the prevention of CHD, cancer and other inflammatory diseases, because other clinical studies clearly indicated the risk of LA for CHD (fig. 5, 9–11, 67; tables 4, 5, 17).

Chapter 8

Mechanisms by Which Dietary Fats Affect Coronary Heart Diesease Mortality

The mechanisms by which dietary FAs affect CHD mortality have not been fully elucidated, particularly with respect to the contribution of saturated and monounsaturated FAs. In this chapter, possible causal mechanisms are described focusing on the important balance of ω3 and ω6 FAs in both diets and tissues.

8.1.

Two Dietary Imbalances Cause Diet-Induced Dyslipidemias That Associate with Impaired Human Health

Many disorders in human health associate with diet-induced dyslipidemias that come from two readily prevented dietary imbalances: imbalanced intake and expenditure of food energy and imbalanced intake of ω6 and ω3 fats.

Fig. 75

Postprandial hypertriglyceridemia and NEFA as factors causing oxidative injury in vascular tissues

Adapted from Lands [2005].

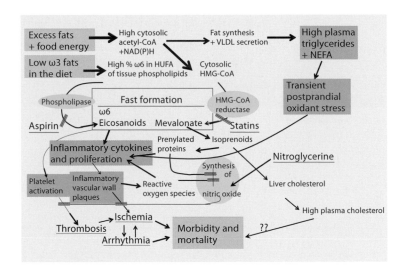

Conclusion: *The circulating TG in plasma rises and falls after every meal (postprandial conditions) in a transient manner – especially after big, high-energy meals. The circulating TG are carried by chylomicrons from the gut and by VLDL secreted by the liver (reflecting more food calories eaten than burned to CO_2 during this transitional time). The circulating TG are cleaved in the vascular system by lipoprotein lipase to give circulating LDL and lots of NEFA that quickly exit the blood and enter vascular tissues.*

Prevention of Coronary Heart Disease

Our comments: The formation of high circulating LDL means that lots of NEFA have been released in the blood stream, but most people don't tell that part of the story. After meals bring extra food to the bloodstream, the liver synthesizes and secretes VLDL into the blood (postprandial lipemia) where lipoprotein lipase converts it into LDL with a release of lots of nonesterified FAs (NEFA) that can aggravate local vascular inflammatory conditions. Transiently elevated NEFA (saturated, monounsaturated or polyunsaturated) can create transient oxidant stress in vascular tissues that is reversible most of the time and leaves no trace of the transitory 'insult' that occurs – especially when people eat lots of vegetables with lots of antioxidants. Saturated FAs seem worse than highly unsaturated FAs in causing longer and more severe inflammatory responses. These events are not caused by the amount of circulating TC, but they partially relate to the amount of circulating TG (that will release NEFA).

Occasionally (perhaps less than 1% of the time), the oxidant stress triggered by NEFA alters local cellular signaling and cellular abundances, creating a small inflammatory site that lingers and becomes a continuing chronic inflammatory site where more serious pathology slowly develops over time. People eat about one thousand meals each year. If 99.9% gave reversible insults and only 0.1% of the meals caused a chronic inflammatory site that means production of one long-term inflammatory vascular site every year. The study tells the rest of this story about the slow steady accumulation of vascular pathobiology in young, middle-aged and elderly Americans that needs preventive nutrition education beginning at adolescence (fig. 45).

Perhaps researchers will gather enough data to show that a high percent of saturated fats in the diet gives NEFA that cause more severe production of inflammatory vascular lesions than unsaturated fats do. So far, the data on types and amounts are still not very clear, although the great attention currently given to LDL should also recognize that it was formed in the blood along with lots of NEFA. The NEFA may be a more important causal mediator of vascular inflammation than the LDL (or the LDL cholesterol!). In the following sections, other potential factors are considered.

Fig. 76

Pronounced postprandial lipemia impairs endothelium-dependent dilation of the brachial artery in men

A part of the original figure was reproduced from Gaenzer et al. [2001], with permission from Elsevier.

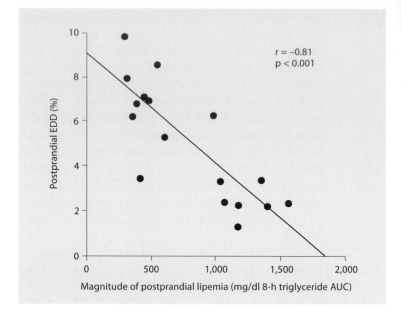

Pronounced postprandial lipemia has been established as a risk factor for cardiovascular disease, but reports regarding its effect on endothelial function have been controversial. In the present study, the influence of a standardized fatty meal with its ensuing postprandial lipemia of highly varying magnitude on endothelium-dependent dilation (EDD) was investigated.

Methods: In 17 healthy, normolipidemic men, EDD of the brachial artery was quantified in two series of three measurements each. In both series initial measurements were performed at 08:00 h after an overnight fast followed by measurements at 12:00 and 16:00 h, in the first series with continued fasting and in the second following the ingestion of a standardized fatty test meal 4 and 8 h postprandially.

Results: Measurements of EDD in the fasting state revealed the recently appreciated diurnal variation with higher values in noon and afternoon hours compared with morning values (2.5 ± 1.6% at 08:00 h, 7.5 ± 2.7% at 12:00 h, and 7.0 ± 2.1% at 16:00 h, $p < 0.001$ by analysis of variance). Postprandial EDD values measured at 12:00 h were, at the average, lower than fasting EDD values measured at 12:00 h and correlated inversely with the magnitude of postprandial triglyceridemia ($r = -0.81$, $p < 0.001$). In multivariate analysis, higher postprandial lipemia was associated with impaired postprandial EDD ($p < 0.001$) independent of fasting TG, LDL-C and HDL-C, insulin, age and body mass index.

Prevention of Coronary Heart Disease

Conclusions: *Pronounced postprandial lipemia is associated with transient impairment of endothelial function. Our findings support the notion that impaired triglyceride metabolic capacity plays an important role in atherogenesis.*

Our comments: EDD expressed by % may indicate the presence of the limited capacity of artery to metabolize TG; greater magnitude of postprandial lipemia may simply mean that a longer time is required to metabolize it. The question remains whether or not the EDD (%) really represents the degree of 'impairment'. The number of subjects (n = 17) needs to be increased to be conclusive.

8.2.

Anti-Atherosclerotic Actions of ω3 Fatty Acids

Fig. 77

Comparison of plasma FAs and urinary eicosanoid metabolites – Inuit vs. Danes

Data taken from Dyerberg et al. [1978], and from Fischer et al. [1986].

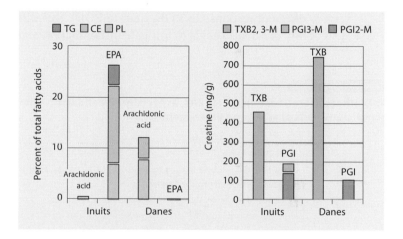

Conclusions: *Unlike AA, EPA does not induce platelet aggregation in human platelet-rich plasma (PRP), probably because of the formation of thromboxane A₃ (TX A₃) which does not have platelet aggregating properties. Moreover, EPA, like AA, can be utilized by the vessel wall to make an anti-aggregating substance, probably a delta-17-prostacyclin (PGI₃). This finding suggests that, in vivo, high levels of EPA and low levels of AA could lead to an antithrombotic state in which an active PGI₃ and a nonactive TXA₃ are formed. Eskimos have high levels of*

EPA and low levels of AA and they also have a low incidence of myocardial infarction and a tendency to bleed. It is possible that dietary enrichment with EPA will protect against thrombosis.

Our comment: The effects of EPA and the EPA/AA ratio on prostanoid (PG) metabolism and function have been reviewed recently by Smith (2005), and are summarized as follows.

One pronounced effect of fish-oil-induced increases in EPA/AA ratios is decreased PG formation from AA via cyclooxygenase-1, because EPA inhibits this isoform. In addition, cells lacking endogenous alkyl-peroxide-generating systems and thus having a low 'peroxide tone' cannot oxygenate EPA via cyclooxygenase-1. Platelets, however, which are equipped with a lipoxygenase that can produce an abundance of hydroperoxide from AA, can form small amounts of thromboxane A_3 from EPA via cyclooxygenase-1. A second major consequence of elevated EPA/AA ratios is significantly increased production of 3-series PGs, including PGE_3, via cyclooxygenase-2. There are four PGE receptor subtypes and at least one of these types – not yet identified – has a significantly different response to PGE_3 than to PGE_2; this difference may underlie the ability of ω3 fatty acids to mitigate inflammation and tumorigenesis.

Table 21

Anti-atherosclerotic actions of ω3 FAs, especially EPA and DHA – Summary of reported beneficial effects of ω3 FAs on atherogenesis

Anti-inflammatory action
Competitive inhibition of the production of ω6 eicosanoids by EPA
Suppression of PAF production by ALA
Suppression of the production of inflammatory cytokines
Potential suppression of C-reactive protein (CRP), an inflammatory marker
Suppression of the expression of cell-adhesion molecules
Antithrombotic action
Decrease in TXA_2 and increase in PGI_3
Improved peripheral blood flow
Decrease in pulse wave velocity
Decrease in blood viscosity
Increase in erythrocyte deformability
Suppression of oxidized LDL formation
Anti-arrhythmic action [Kang and Leaf, 2000]
Regulation of gene expression (common to HUFAs regardless of ω6 or ω3 type)
A part of genes regulated by HUFA were summarized in figure 80

Kang and Leaf [2000] reviewed anti-arrhythmic actions of ω3 FAs as follows:

' If the arrhythmias were first induced, adding the EPA to the superfusate terminated the arrhythmias. This anti-arrhythmic action occurred with dietary ω3 and ω6 PUFAs; saturated fatty acids and the monounsaturated oleic acid induced no such action. Arachidonic acid (AA; 20:4n-6) is anomalous because in one-third of the tests it provoked severe arrhythmias, which were found to result from cyclooxygenase metabolites of AA. When cyclooxygenase inhibitors were added with the AA, the antiarrhythmic effect was like those of EPA and DHA. The action of the ω3 and ω6 PUFAs is to stabilize electrically every myocyte in the heart by increasing the electrical stimulus required to elicit an action potential by approximately 50% and prolonging the relative refractory time by approximately 150%. These electrophysiologic effects result from an action of the free PUFAs to modulate sodium and calcium currents in the myocytes. The PUFAs also modulate sodium and calcium channels and have anticonvulsant activity in brain cells. '

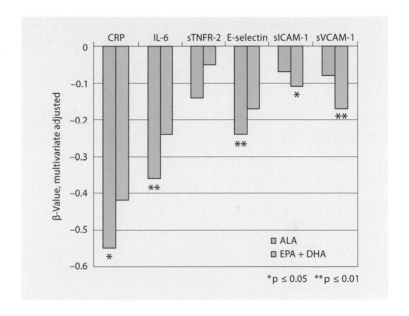

Fig. 78

Consumption of ω3 FAs is related to plasma biomarkers of inflammation and endothelial activation in women – Nurses' Health Study Cohort

Data taken from Lopez-Garcia et al. [2004].

A cross-sectional study was conducted for 727 women from the Nurses' Health Study I cohort, aged 43–69 years, apparently healthy at the time of a blood draw in 1990. Dietary intake was assessed by a validated FFQ in 1986 and 1990. CRP levels were 29% lower among those in the highest quintile of total ω3 FAs, compared with the lowest quintile; IL-6 levels were 23% lower, E-selectin levels 10% lower, soluble intracellular adhesion molecule (sICAM-1) levels 7% lower, and soluble vascular adhesion molecule (sVCAM-1) levels 8% lower. The intake of ALA was inversely related to plasma concentrations of

CRP ($\beta = -0.55$, p = 0.02), Il-6 ($\beta = -0.36$, p = 0.01), and E-selectin ($\beta = -0.24$, p = 0.008) after controlling for age, BMI, physical activity, smoking status, alcohol consumption, and intake of LA (ω6) and saturated fat. Long-chain ω3 FAs (EPA and DHA) were inversely related to sICAM-1 ($\beta = -0.11$, p = 0.03) and sVCAM-1 ($\beta = -0.17$, p = 0.003). Total ω3 FAs had an inverse relation with CRP ($\beta = -0.44$, p = 0.007), IL-6 ($\beta = -0.26$, p = 0.009), E-selectin ($\beta = -0.17$, p = 0.004), sICAM-1 ($\beta = -0.07$, p = 0.02), and sVCAM-1 ($\beta = -0.10$, p = 0.004). These associations were not modified by intake of vitamin E, dietary fiber, *trans*-FAs, or by the use of postmenopausal hormone therapy.

Conclusion: *This study suggests that dietary ω3 FAs are associated with levels of these biomarkers reflecting lower levels of inflammation and endothelial activation, which might explain in part the effect of these FAs in preventing cardiovascular disease.*

Our comments: Both PUFA (ALA) and HUFA (EPA, DHA) of ω3 type were effective in lowering biomarker levels of inflammation and endothelial activation.

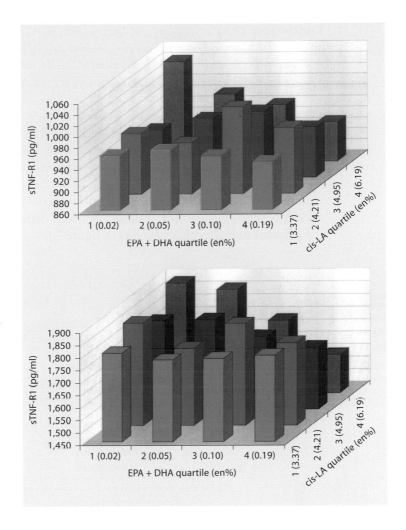

This figure was reproduced from Pischon et al. [2003], with permission from Lippincott Williams & Wilkins.

Fig. 79

Habitual dietary intake of ω3 and ω6 FAs in relation to inflammatory markers among US men and women

Habitual dietary ω3 FA intake and its interaction with ω6 FAs were investigated in relation to plasma inflammatory markers among 405 healthy men and 454 healthy women. Intake of the ω3 FAs EPA and DHA was inversely associated with plasma levels of sTNF-R1 and sTNF-R2. Little if any association was found between ω3 FA intake and tumor necrosis factor receptors among participants with low intake of ω6 but a strong inverse association among those with high ω6 intake.

Conclusion: *These results suggest that ω6 FAs do not inhibit the anti-inflammatory effects of ω3 FAs and that the combination of both types of FAs is associated with the lowest levels of inflammation. The inhibition of inflammatory cytokines may be one of the observed beneficial effects of these FAs on chronic inflammatory-related diseases.*

Our comments: Effectiveness of ω3 FAs (EPA, DHA) in lowering the levels of biomarkers of inflammation was demonstrated. The interaction between ω6 and ω3 FAs appears to be complex, but the competitive aspects are clearly seen at the highest LA intake group. Dietary LA (ω6) in large amounts is not hypocholesterolemic in the long run, raises inflammatory lipid mediators and the incidence of related diseases. It should be noted that the level of sTNF-R2 was not affected significantly by ω3 FAs in figure 78.

Fig. 80

Differential effects of various FAs on gene expression through SREBP and PPAR

Sampath and Ntambi [2004] and Clarke [2004].

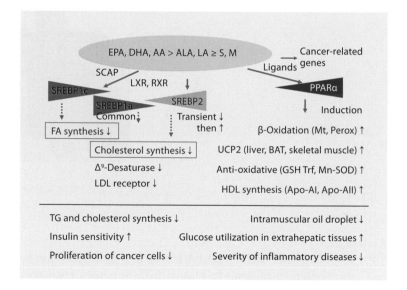

Our interpretation on currently available data: Gene expressions mediated by two transcriptional elements (SREBP and PPARα) appear to be affected by the chain length and number of double bond (melting point) of the FAs. Compared to saturated (S) and monounsaturated (M) FAs, unsaturated FAs with 20 and 22 carbons (EPA, DHA, AA) induce enzymes of β-oxidation in mitochondria and peroxisomes, and increase an uncoupling protein by serving as ligands for PPARα, resulting in the stimulation of thermogenesis. TG molecular species containing FAs with very low melting points (EPA, DHA, AA) dis-

Prevention of Coronary Heart Disease

solve surrounding phospholipids at body temperature and appear to be inadequate as depot fats. Induction of peroxisomal β-oxidation enzymes and UCP may form a part of mammals' adaptation to cold climate. On the other hand, these unsaturated FAs suppress the expression of SREBP and cholesterol synthesis compared to S, M and LA.

LA suppresses the syntheses of FAs and cholesterol compared with S and M but this effect is much weaker than EPA and DHA. Similarly, ALA (α-linolenic acid) proliferates peroxisomes but this activity is much lower than fish oil EPA and DHA, and LA is essentially ineffective for peroxisomal proliferation. These regulations of gene expression mediated by transcriptional elements mainly depend on the chain length and number of double bond of FA, regardless of whether the FA is ω3 or ω6.

The ω6/ω3 balance of dietary FAs affects gene expressions related to apoptosis, cell-proliferation and even to brain functions (not included in this figure). Minor unidentified factors in vegetable oils also affect gene expression in different ways (chap. 8.4.). This field of research is in rapid progress and our interpretation of the differential effects of FAs on gene expression is tentative. The amounts of fats and oils, fatty acid compositions, feeding periods, energy balance and other nutritional factors would affect expressions of these genes but in vivo data available are not enough to summarize the features.

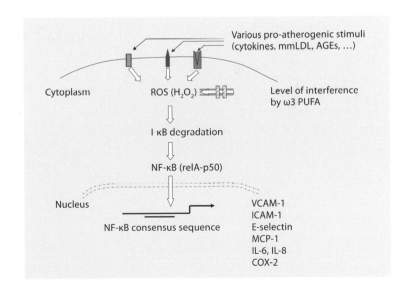

Fig. 81

ω3 FAs and the regulation of expression of endothelial pro-atherogenic and pro-inflammatory genes

Reproduced from De Caterina and Massaro [2005], with kind permission from Springer Science and Business Media.

Authors' explanation: *A model of the putative site of action of ω3 PUFA, the most potent of which appears to be DHA, in inhibiting endothelial activation, thus potentially decreasing early atherogenesis. ω3 PUFA would act downstream to receptors for various atherogenic stimuli (such as minimally modified LDL (mmLDL), the advanced glycation endproducts (AGEs) or inflammatory cytokines), likely at the level of reactive oxygen species (ROS), the most relevant of which appears to be hydrogen peroxide (H_2O_2). This would activate the nuclear factor-κB (NF-κB) system of transcription factors, likely through the activation of the degradation of the inhibitor I-κB, allowing the free active NF-κB heterodimers (rel A-p50) to translocate into the nucleus, bind to specific consensus sequences in a number of NF-κB-responsive genes (including genes for vascular cell adhesion molecule-1 (VCAM-1), intercellular adhesion molecule-1 (ICAM-1), E-selectin, monocyte chemoattractant protein-1 (MCP-1), interleukin (IL)-6 and IL-8, and cyclooxygenase-2 (COX-2)), driving their increased transcription.*

Our comment: Endothelial activation is an early step of atherogenesis involving up-regulation of pro-inflammatory and pro-atherogenic genes, and the level of ω3 PUFA action was elucidated here. Although DHA was most active among major ω3 PUFAs under the conditions, their relative activity in vivo may vary depending on the period of dietary manipulation, the amounts of other competitive FAs and the status of energy balance, because DHA is known to proliferate peroxisomes that β-oxidize various FAs at differential rates.

Prevention of Coronary Heart Disease

8.3.

The Injury-Inflammation-Ischemia Hypothesis of Atherosclerosis

Fig. 82

Injury-inflammation-ischemia hypothesis of atherosclerosis

Adapted from Okuyama et al. [2000].

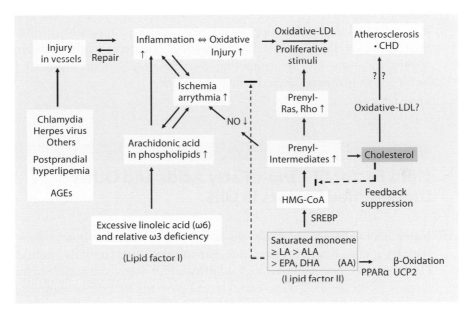

Chlamydia, herpes virus and other pathogens are presumed to injure blood vessels [Ross, 1999]. Postprandial hypertriglyceridemia is also proposed to trigger the vessel injury (fig. 75, 76). Advanced glycation endproducts (AGE) may also trigger endothelial injury (fig. 81). Inflammatory reactions would follow to repair the injury. When LA intake is high and membrane phospholipids are saturated with AA, the enhanced LA cascade leads to persistent inflammation by over- and unbalanced production of eicosanoids (elevated TXA_2/PGI_2 ratio and increased leukotriene production). Increased thrombotic tendency leads to ischemia and inflammation. Reactive oxygen species produced by hypoxic (ischemic) mitochondria and/or inflammatory cells attack LDL to form oxidized LDL, stimulate cell-proliferation and thereby accelerate atherogenesis. The lipid factor I in this figure is associated with ω6/ω3 balance and the LA cascade.

On the other hand, cholesterol synthesis mediated through SREBPs is stimulated by FAs; the stimulatory activity appears to be roughly in the order of S, M ≥ LA > ALA > EPA, DHA (AA), an interpretation from experiments using cultured cells and animals (Lipid Factor II). At this step, relatively little difference is observed neither between animal fat (S, M) and high-LA vegetable oils, nor between ω6 and ω3 FAs. Thus, S and M, and LA to a slightly lesser extent, possibly elevate the levels of isoprenyl intermediates, suppress the formation of vasodilatory NO, promote cell proliferation through prenylation of oncogene products (Ras, Rho) and accelerate atherogenesis. High levels of TC and oxidized cholesterol feedback suppress cholesterol synthesis. Thus, ω3 FAs, especially EPA and DHA, suppress atherogenesis by competitively inhibiting the LA cascade (lipid factor I), by suppressing cholesterol synthesis (prenyl intermediate levels) and by promoting β-oxidation and thermogenesis (lipid factor II), all involving altered gene expressions.

8.4.

Lipid Factor III – *trans*-Fatty Acids and Other Unidentified Factors in Oils

Nutritional values of vegetable oils have been evaluated primarily based on their FA composition. However, we began to realize that factors other than FAs (TG) influence animal physiology significantly at doses comparable to human intakes. Some kinds of vegetable oils and partially hydrogenated oils shortened the survival of stroke-prone SHR (SHRSP) rats unusually as compared with others. The free FA fractions derived from canola oil and hydrogenated-soybean oil exhibited no or significantly reduced activities, indicating the presence of minor components other than FAs (TG) [Huang et al., 1997; Ratnayake et al., 2000].

On the other hand, *trans*-FAs in hydrogenated oils have been regarded as a risk factor of CHD (related to elevated cholesterol levels), and movements to exclude them from our food environment are rapidly in progress, although epidemiological evidence is not clear enough as summarized in the earlier part of this section. Here, we take into account the possible involvement of factors other than phytosterols and *trans*-FAs that influence animal physiology. Those vegetable oils with antinutritional factors cause kidney lesion, decrease megakaryocyte counts in the bone marrow, reduce blood platelet counts and shorten survival in SHRSP rats [Okuyama et al., 1996; Ohara et al., 2005]. Decreased platelet number by canola oil-containing milk replacer is detected in piglets as well [Innis et al., 1999]. Their potential implication in human nutrition is briefly summarized in the latter part of this section.

Fig. 83

Intake of FAs and Risk of CHD in a cohort of Finnish men – Data for *trans*-FA

Data taken from Pietinen et al. [1997].

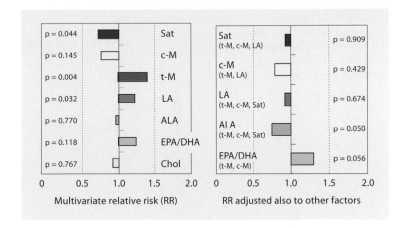

Male smokers (n = 21,930) aged 50–69 years were followed for 6.1 years. Major coronary events (n = 1,399) and coronary deaths (n = 635) were noted. After controlling for age, supplement group, several coronary risk factors, total energy and fiber intake, the authors observed a significant positive association between *trans*-FA intake or LA intake and the risk of coronary deaths. In the multivariate model, the intakes of *trans*-FA and ω3 FAs from fish were directly related to the risk of coronary deaths. There was no association between intakes of saturated or *cis*-monounsaturated FAs, LA or ALA, or dietary cholesterol and the risk of coronary deaths. All the associations were similar but somewhat weaker for all major coronary events.

Concluding remarks: *The selective nature of this cohort (middle-aged, smoking men eating a diet high in fat) warrants relatively cautious extrapolation to other populations.*

Our comments: The intake of LA was positively associated with the risk of coronary events, but the association became non-significant when adjusted to other FAs. The intake of *trans*-FAs was highly positively associated with risk in a multivariate analysis (p = 0.004) but no data were given for its association after adjustment to other FAs. When adjusted to other FAs, ALA (ω3) intake was inversely and fish ω3 FA intake was positively associated with the risk of coronary events. The authors discussed the problems of mercury pollution of local freshwater fish in this area.

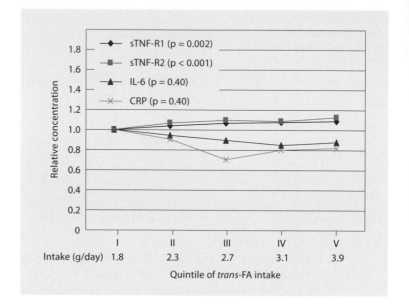

Fig. 84

Dietary intake of *trans*-FAs and systemic inflammation in women – From Nurses' Health Study I and II

Data taken from Mozaffarian et al. [2004].

In 823 generally healthy women in the Nurses' Health Study I and II, concentrations of soluble tumor necrosis factor-α receptors 1 and 2 (sTNF-R1, sTNF-R2), interleukin-6 (IL-6), and C-reactive protein (CRP) were measured. Usual dietary intakes assessed from 2 semiquantitative food-frequency questionnaires were averaged for each subject.

Results: In age-adjusted analyses, *trans*-FA intake was positively associated with sTNF-R1 and sTNF-R2 (p for trend < 0.001 for each): sTNF-R1 and sTNF-R2 concentrations were 10% (+108 pg/ml) and 12% (+258 pg/ml) higher, respectively, in the highest intake quintile than in the lowest. These associations were not appreciably altered by adjustment for body mass index, smoking, physical activity, aspirin and NSAID use, alcohol consumption, and intakes of saturated fat, protein, ω6 and ω3 FAs, fiber, and total energy. Adjustment for serum lipid concentrations partly attenuated these associations, which suggests that they may be partly mediated by effects of *trans*-FA on serum lipids. *trans*-FA intake was not associated with IL-6 or CRP concentrations overall but was positively associated with IL-6 and CRP in women with higher body mass index (p for interaction = 0.03 for each).

Prevention of Coronary Heart Disease

Conclusions: *trans-FA intake is positively associated with markers of systemic inflammation in women. Further investigation of the influences of trans-FAs on inflammation and of implications for coronary disease, diabetes, and other conditions is warranted.*

Our comment: Despite the deduced conclusion and high degree of statistical significance (very low p values), the data shown in this figure are not convincing enough; relative marker concentrations (the highest quintile to the lowest quintile) of 1.09 (sTNF-R1) and 1.12 (sTNF-R2) were significant (p < 0.002) but those of 0.87 (IL-6) and 0.82 (CRP) were not (p = 0.40).

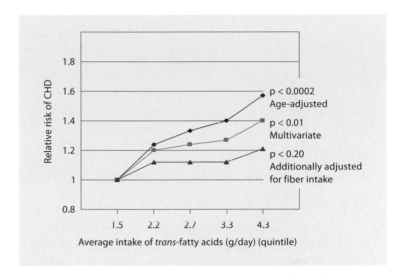

Fig. 85

Relative risk of total myocardial infarction according to quintiles of dietary S, cholesterol, LA, ALA and Keys' score – Data for *trans*-FAs

Data taken from Health Professionals Follow-Up Study in the USA [Ascherio et al., 1996].

Health professionals (45–75 years old, n = 43,757) were followed for 6 years from 1986. Figures are relative risks. Details are described in the legend to table 3.

Conclusion: *These data do not confirm the strong association between intake of saturated fat and risk of CHD seen in international comparisons. However, they do fit the hypothesis that saturated fat and cholesterol intakes affect the risk of CHD as predicted by their effects on TC concentration. They also support a specific preventive effect of ALA intake.*

Our comments: Among health professionals in the USA, *trans*-FA intake was apparently positively associated with MI. However, the association became weaker after multivariate analysis, and insignificant when adjusted further to dietary fiber intake.

Fig. 86

Dietary fat intake and risk of CHD in women – *trans*-FA and LA

Data taken from 20 Years of Follow-up of the Nurses' Health Study [Oh et al., 2005].

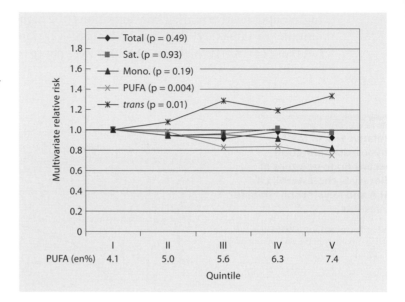

The association of dietary fat and specific types of fat with risk of CHD was examined among 78,778 women initially free of CVD and diabetes in 1980. The association between intakes of PUFAs and *trans*-fat with CHD risk was most evident among women younger than age 65 years. Multivariate relative risk was adjusted for age, body mass index, smoking, alcohol intake, parental history of MI, history of hypertension, menopausal status, hormone use, multivitamin use, vitamin E supplement use, physical activity, and energy, protein and cholesterol intake, and intakes of saturated (Sat), monounsaturated (Mono), polyunsaturated (PUFA), *trans*-fat (Trans); ALA; marine ω3 FAs; cereal fiber and fruits and vegetables.

Conclusion: *Findings continue to support an inverse relation between polyunsaturated fat intake and CHD risk, particularly among younger women. In addition,* trans-*fat intake was associated with increased risk of CHD, particularly for younger women.*

Our comments: The relative risk of *trans*-fat was 1.33 (p = 0.01) in this study compared with 1.21 (p = 0.20) in the Health Professionals Study (fig. 87). Highly positive association between *trans*-FA and CHD has been noted in Costa Rican adults for whom relative risks were 2.94 (all *trans*-FAs) and 5.05 (18:2 *trans*-FAs) [Baylin et al., 2003], but not in the industrialized countries. We point out the problems of statistical treatments; the choice of factors for adjustments may influence the outcome greatly.

As to the effects of dietary LA on CHD, the relative risk of PUFA (mostly LA) for the highest vs. the lowest quintile was 0.75 (p = 0.004) in the Nurses' Health Study while it was 1.04 (p = 0.89) in the Health Professional Study (table 3; fig. 87), which is far from the value to account for severalfold difference in CHD mortalities among different populations (e.g. fig. 12, Seven Countries Study). Although a protective effect of LA in CHD mortality has been continuously shown by this group, it is difficult to accept that increasing the intake of LA is beneficial for the prevention of atherosclerosis. The LA may be a surrogate marker of vegetable oil containing anti-thrombotic (apoplexy-accelerating) factors leading to decreased platelet number. One of such factors in hydrogenated oils may be dihydro-vitamin K_1 and related compounds (fig. 88).

Fig. 87.

Relative risk of total myocardial infarction according to quintiles of dietary S, LA and ALA

Data taken from Health Professionals Follow-Up Study in the USA [Ascherio et al., 1996].

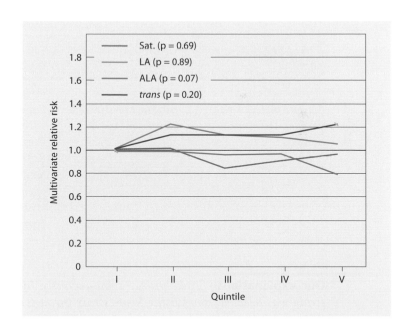

A part of table 3 (multivariate analysis adjusted for dietary fiber) was shown here to compare with figure 86. See details in table 3.

Our comments: Two epidemiological studies performed in the USA (fig. 86, 87) differ in gender proportion of the selected subjects, but adjusted relative risks shown in these figures are relatively similar. However, the conclusions derived from these two studies are quite different because of the difference in p values. One factor to be considered is a possible difference in the degrees of health consciousness; nurses may have paid special attention to energy balance, physical exercise, choice of fats and oils and other potential protective factors compared with health professionals, which brought about a small difference in the outcome. An important fatty acid, ALA, is not included in figure 86. Regardless of the difference in conclusions from these studies, the impact of dietary LA and *trans*-FAs on CHD is very small, if any, compared with several-fold difference in CHD mortality among different populations (fig.12–15).

Table 22

Nutritional values of fats and oils estimated by the survival rates of SHRSP rats

Data taken from Okuyama et al. [1996] and Ratnayake et al. [2000].

Fats and oils that extend survival	Control oils	Fats and oils that shorten survival unusually
Perilla seed oil (ALA-rich)	Soybean oil	Canola
Linseed oil (ALA-rich)	Safflower oil (LA-rich)	Rapeseed oil (erucic acid-rich)
Fish oil (DHA-rich)	Sesame oil (LA-rich)	Olive oil
Butter		Corn oil
Lard		Safflower oil (oleic acid-rich)
		Sunflower oil (oleic acid-rich)
		Evening primrose oil
		Hydrogenated soybean oil
		Hydrogenated canola oil

Our explanations: *Stroke-prone, spontaneously hypertensive (SHRSP) rat strain was selected from Wistar/Kyoto strain through SHR strain, and develops hypertension and dies of cerebral bleeding in high frequency when NaCl solution was loaded as drinking water. Diets supplemented with 10% fat or oil were fed* *to SHRSP rats from weaning, and survival rates were estimated. The data combined from those of different laboratories and from different sets of experiments are classified into three groups as summarized above. It should be noted that all fats and oils were not estimated under the same conditions; hence this grouping is a rough measure of their effects on survival of SHRSP rats. Difference in mean survival times of soybean oil group and canola oil group was more or less 40% in the absence of NaCl loading.*

Our comments: As the free fatty acid fractions derived from canola oil and hydrogenated soybean oil exhibited no or significantly reduced activity, the difference in the observed survival times was not accounted for by the difference in their fatty acid and phytosterol compositions. Efforts to identify the presumed minor components are in progress, but one of the factors in hydrogenated oils is likely to be dihydro-vitamin K_1 (fig. 88).

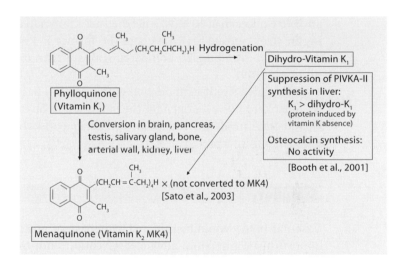

Fig. 88

An antinutritional factor in hydrogenated vegetable oils, and unidentified nutritional factors in some vegetable oils

Booth et al. [2001], Sato et al. [2003] and Okuyama et al. [1998].

Our explanations: Vitamin K_1 with one double bond in the four isoprenoid side chains is metabolized in our body to vitamin K_2 (menaquinone 4 with four double bonds in the side chain), which appears to have physiological activities different from that of vitamin K_1. Vitamin K_1 is rich in vegetable oils such as soybean oil, canola oil and olive oil, and dihydro-vitamin K_1 is produced by hydrogenation, which is partially active for the formation of mature coagulation factors, but is ineffective for the production of osteocalcin involved in bone metabolism [Booth et al., 2001]. Matrix Gla (carboxylated glutamyl) proteins are also produced by vitamin K-dependent enzymes and play roles in cellular proliferation. Inhibition of vitamin K-dependent reactions could cause decreased thrombotic tendency, increased bleeding tendency and altered physiology of other unraveling functions of different vitamin Ks. In this sense, dihydro-vitamin K_1 at adequate doses may be beneficial for the prevention of thrombotic diseases but anti-nutritional at higher doses.

Another reason to consider these unidentified factors in vegetable oils, exists in the periods before changes in dietary fats and oils exert significant effects on CHD in clinical trials. In the Helsinki Businessmen Study (fig. 10; table 4), the difference in the CHD mortalities of the control and intervention groups became clearer and greater after 10 years whereas in the Lyon Diet Heart Study the choice of canola and olive oil in the intervention group exerted significant preventive effects within 2 years (fig. 67). In spite of the 70% risk reduction within a few years found in the Lyon Diet Heart Study it appears to be too early to ascribe the observed effects to reduced intake of LA and increased intake of ALA and oleic acid. These oils may contain additional factors other than TG (FAs) that are anti-thrombotic at adequate doses as canola oil and olive oil, each at 10% (w/w) of diet, are known to accelerate cerebral bleeding in stroke-prone SHR rats, and the former is known to decrease platelet counts in rats and piglets compared with soybean oil [Ohara et al., 2005; Innis et al., 1999].

Although a significant portion of this section needs further confirmation, we must keep our eyes open for the potential involvement of dihydro-vitamin K_1 and other unidentified factors in vegetable oils when dietary FAs-CHD relationship is interpreted.

Summary

Vascular injury could be initiated by persistent postprandial hyperlipidemia, opportunistic infection, advanced glycation endproducts and/or by others. Vascular injury and repair processes involve inflammation, and enhanced LA (AA) cascade activity can lead to persistent vascular inflammation leading to ischemia. Reactive oxygen species (ROS) are produced from inflammatory cells to oxidize LDL, and initiate inflammatory and atherogenic processes through transcriptional factors (e.g. NF-κB) in vascular cells. We postulate three lipid factors involved in these processes. Lipid factor I is related to $\omega6/\omega3$ balance of FAs and lipid factor II is related to the chain length and degree of unsaturation of FAs affecting gene expression of lipolysis and lipogenesis. Greenland natives and Mediterranean people have very low CHD mortality rates compared with people in the USA and UK, although they eat relatively large amounts of saturated and monounsaturated fatty acids (lipid factor II), indicating that low ω6/ω3 ratios (lipid factor I) may overcome the impact of large amounts of saturated and monounsaturated fatty acids (lipid factor II) on CHD. The extent to which *trans*-FAs serve as a risk factor of CHD in industrialized countries remains to be evaluated more accurately, but we must keep in mind the possible involvement of other minor components in vegetable oils and hydrogenated oils in the dietary FAs-CHD relationship (lipid factor III).

Chapter 9

Cancers Common in the USA Are Stimulated by ω6 Fatty Acids and Large Amounts of Animal Fats, but Suppressed by ω3 Fatty Acids and Cholesterol

Long-term follow-up studies on general populations performed in different countries revealed that cancer mortality and/or all-cause mortality tended to be lower in higher TC groups, e.g. ≥50 years of age in Austria, ≥60 years of age in France, ≥85 years of age in the Netherlands, ≥71 years of age in the USA and ≥40 years of age in Japan. On the other hand, suppression of the LA (AA) cascade by drugs, dietary manipulation or genetic modification was revealed to be effective in suppressing carcinogenesis. Thus, the LA cascade appears to be involved in both atherogenesis and carcinogenesis. In this chapter, our current understandings on the effects of FAs and cholesterol on carcinogenesis are briefly summarized.

9.1.

Enhanced Linoleic Acid Cascade as a Major Risk Factor of Cancers Common in Industrialized Countries

Fig. 89

Trends of FA intake and mortality from cancers – Japan vs. USA

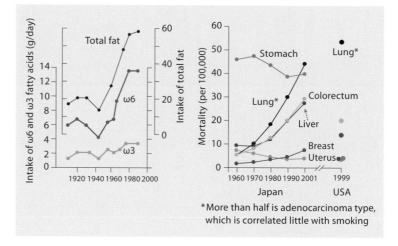

*More than half is adenocarcinoma type, which is correlated little with smoking

Our explanation: Increase in the mortalities from cancers of the lung, mammary gland, colorectum, pancreas, prostate and others preceded in the USA, followed by Japan and other Asian countries over the past several decades (USA-type cancer). Most of them seem to be adenocarcinoma. In Japan, the intake of LA increased 2.5-fold during 1960–1975, reaching a plateau level thereafter, and the increase in cancer mortality followed. The mortality of colorectal cancer in Japan recently exceeded that in the USA, and those of lung, mammary glands and other cancers are sharply increasing. Lung cancer that is increasing in Japan nowadays is mainly of adenocarcinoma type (>50%), which is correlated relatively little with smoking but is shown in animal experiments to be stimulated by dietary LA.

Conclusions: *Mortality from cancers in Japan comprises >30% of all-cause deaths. Because CHD is the leading cause of death in the USA but is kept low in Japan, the mortalities from other cancers in Japan will also exceed those of the USA if the trends continue.*

Fig. 90

Cancer researches from different fields now focus on the LA cascade

Adapted from Okuyama [2000, 2003].

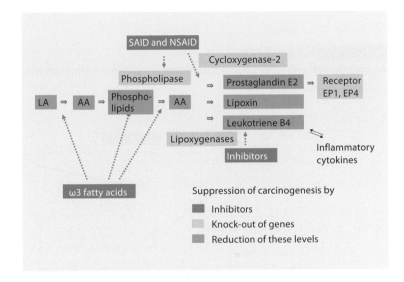

Our explanation: In animal experiments, high-LA vegetable oils are shown to stimulate but oils low in ω6/ω3 ratios (perilla seed oil, linseed oil, fish oil) suppress carcinogenesis (nutritional approach). The ω3 FAs are known to exert their anticarcinogenic activity mainly by competitively inhibiting the LA cascade. Steroidal and nonsteroidal anti-inflammatory drugs as well as 5-lipoxygenase inhibitors suppress the LA cascade and thereby suppress carcinogenesis (pharmaceutical approach). Knock-out of genes related to the LA cascade is also effective in suppressing carcinogenesis, e.g. genes for cytosolic phospholipase A2 to release AA, cyclooxygenase 2 to produce eicosanoids and prostaglandin E-receptors (EP1, EP4) (gene technological approach). Effectiveness of all these approaches clearly indicates that suppressing the LA cascade is useful for the prevention of cancers of the USA type.

Cancers Common in the USA Are Stimulated by ω6 FAs and Large
Amounts of Animal Fats, but Suppressed by ω3 FAs and Cholesterol

145

Our comments: Increase in LA intake in the past several decades is a major risk factor for many types of cancers. However, epidemiological studies revealed that both ALA (ω3) intake and serum ALA level were positively associated with prostate cancer incidence. In this case, ALA is likely to be a surrogate marker of cancer-promoting factor(s) in vegetable oils, because (a) the serum level of ALA is higher but the incidence of prostate cancer is lower in Japan than in the USA, and (b) high-ALA perilla oil suppressed carcinogenesis in the prostate compared with high-LA safflower oil in experimental animals [Mori et al., 2001]. Thus, ALA is unlikely to promote carcinogenesis in the prostate. Unidentified factors in vegetable oils may also be involved in the high incidence of prostate and lung cancer (chap. 8.4.).

Fig. 91

FAs and cholesterol affecting carcinogenesis – A hypothetical scheme

Adapted from Okuyama [2003].

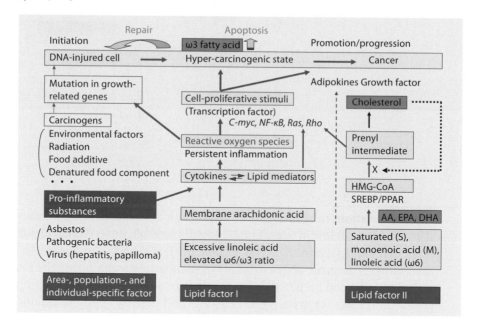

Our Explanation: Much effort has been made to seek environmental mutagens that injure genes leading to the initiation of carcinogenesis. However, the impact of environmental carcinogens may be much less than what we presumed before, because even one of the most potent carcinogens, e.g. amino acid de-

Prevention of Coronary Heart Disease

rivative in broiled meats and fish, requires huge amounts of broiled, baked or deep-fried foods to cause cancer when the carcinogenic doses in animal experiments were applied to the human food environment. On the other hand, persistent inflammation has been recognized in clinical fields to lead to carcinogenesis. For example, hepatitis virus causes hepatic inflammation and chronic hepatitis often leads to hepatic cancer. Asbestos and glass beads do not easily dissolve in animal bodies when administered, hence would not injure genes directly. However, these foreign substances cause inflammation, inflammatory cells are recruited and tumors are formed surrounding these substances. Stomach cancer could well be correlated with chronic *Helicobacter pylori* infection. After World War II in Japan, both mortalities from tuberculosis and pneumonia decreased rapidly but only the age-adjusted mortality from pneumonia began to increase after the 1970s, and the mortality from lung cancer followed. Subjects suffering from ulcerative colitis and Crohn's disease began to increase and so did colorectal cancer. The impact of these inflammation-related factors is different among different populations and among individuals (area, population and individual-specific factor). For example, high incidence of hepatic cancer in Japan and Asia is likely to be correlated with hepatic virus prevailing in this area.

Persistent inflammation is a major risk factor for cancers of the USA type, and enhanced LA cascade leads to overproduction of inflammatory lipid mediators. Lipid mediators and inflammatory peptide mediators possibly form an amplification cascade of inflammation, and lipid mediators (e.g. leukotrienes) upregulate transcription factors (e.g. NF-κB), activate oncogene products (e.g. c-Myc) and stimulate cell proliferation. Lipoxygenase-mediated inflammatory events similar to those causing intima-media thickening of atherosclerosis noted in chapter 5 may be partially suppressed by the negative regulatory region in the 5 LOX promoter of Caucasians, and mutations in this regulatory region may facilitate lipoxygenase expression and tumor proliferation [Goodman et al., 2004]. Reactive oxygen species (ROS) produced from inflammatory cells attack genes to initiate carcinogenesis. In addition, ROS serve as proliferative stimuli leading to carcinogenesis (lipid factor I related). The opposing outcomes from ω6 and ω3 FAs emphasize the need for careful interpretation of how dietary choices affect the proliferation of tumors promoted by inflammatory conditions, conditions especially common in the USA.

High rates of mevalonate and cholesterol synthesis may enhance formation and action of oncogene products such as Ras and Rho, which are activated by isoprenylation. Such increased levels of isoprenyl intermediates would increase this process to form cell-proliferative stimuli. In this sense, animal fats, and LA to a lesser extent, stimulate hepatic cholesterol synthesis and possibly carcinogenesis. Consistently, statins have been shown to suppress carcinogen-induced tumorigenesis. Statins were also carcinogenic by themselves (tables 11, 12; fig. 41, 42), and these contradictory effects are likely to be correlated with divergent functions of isoprenyl intermediates (fig. 40). Impaired synthesis of isopentenyl-adenine, a minor component of tRNA, would disturb protein synthesis, and impaired heme and Coenzyme Q synthesis would cause ischemia and inflammation, a risk factor for carcinogenesis.

Among general populations in Japan and Korea, and among aged populations in the Western countries, the TC value was inversely associated with cancer mortality (fig. 24, 28, 30–34). In peripheral tissues, high TC and high cholesterol availability lead to feedback suppression of cholesterol synthesis and thereby suppression of oncogene activation through Ras and Rho. This may form a part of mechanisms by which high TC is correlated with low cancer mortality. In fact, a diet enriched with cholesterol is shown to suppress carcinogenesis in animal experiments. Other mechanisms would also be involved in the TC-cancer relationship.

Antinutritional actions were observed with some vegetable oils (lipid factor III, chap. 8.4.). Mutagenic activity has been detected in heated cooking oil, which is correlated with a high incidence of lung cancers among nonsmoking women working in Chinese restaurants [Shields et al., 1995]. Thus, unidentified factors (chap. 8.4.) may also be involved.

9.2.

Failure to Reveal the Causal Relationship between Linoleic Acid Intake and Cancer Mortality

In spite of the presence of many lines of evidence from animal experiments to indicate the causal relationship between ω6 FAs and cancer of the USA type, epidemiological studies are not clear enough; some are supportive but others are against the above interpretation (fig. 90, 91). All the considerations presented in chapters 7 and 8 apply to this apparent discrepancy, and no intervention trials are recommended to increase the intake of ω6 fatty acids before the data related to figures 90 and 91 are rationally interpreted.

Prevention of Coronary Heart Disease

Summary

The causal relationship between the enhanced LA cascade and carcinogenesis was well established in animal experiments because suppression of the LA cascade by inhibitors (e.g. NSAID drugs), genetic manipulation to knockout genes related to the LA cascade or ω3 fats resulted in suppression of cancer development. Many reports suggest that cellular signals associated with inflammation increase the proliferation (promotion) of tumors. Some human epidemiological studies involve combined risks for initiation and promotion of cancers and thus give results not consistent with this interpretation, but other epidemiological studies clearly show positive associations of high AA (arachidonic acid) intake and cancer mortality. On the other hand, a high TC value is associated with low cancer mortality in general populations, and effective feedback suppression of isoprenoid intermediates was suggested as a possible mechanism. We emphasize that lipid nutrition based on the 'Cholesterol Hypothesis' was also risky for many types of cancers.

Chapter 10

New Directions of Lipid Nutrition for the Primary and Secondary Prevention of Coronary Heart Disease and Other Late-Onset Diseases

The data summarized in this small book describe the consequences of some metabolic processes that link the foods that we choose to eat with pathology that leads to death. The overall relationship was summarized in figure 75 which indicates two easily corrected imbalances in food intakes that cause diet-induced dyslipidemias that mediate fatal pathophysiological mechanisms. Other lipid factors are also presented in figure 91.

Three pharmaceutical treatments are used widely to moderate the diet-induced CHD problems: aspirin (slowing excessive ω6 signaling), statins (slowing excessive mevalonate formation) and nitroglycerin (supplementing inadequate NO formation). As medical research gives information on how these drugs help to prevent death, it also gives insight into the metabolic molecular mechanisms that were made worse by dyslipidemias caused by imbalanced choices of dietary fats. Dyslipidemias of elevated cholesterol and elevated TG are traditionally discussed without acknowledging the mechanisms that make the biomarkers of HUFA balance and inflammation more relevant. Many disorders in human health associate with diet-induced dyslipidemias that come from two readily prevented dietary imbalances: imbalanced intake and expenditure of food energy and imbalanced intake of ω6 and ω3 fats.

Diet recommendations currently in use should be examined with these dyslipidemias and successful therapeutic tactics in mind. The public is served best by nutritional advice which prevents the diet-induced dyslipidemias that medicines are used to treat.

10.1.

Critical Comments on Adult Treatment Panel III from National Heart, Lung and Blood Institute, NIH, USA

Despite the failure of the MRFIT Study to prove the usefulness of dietary recommendations based on the 'Cholesterol Hypothesis', similar recommendations have been issued from the Adult Treatment Panel III (ATPIII). Japanese medical societies as well as those in other countries have accepted the recommendations and tried to apply them to people. However, the intended hypocholesterolemic effect of raising the P/S ratio and reducing cholesterol intake is transient. The dietary recommendations were ineffective after several years and CHD mortality was unchanged or even increased in a subgroup by such recommendations (MRFIT Study, Helsinki Businessmen Study). Now we know that there is no reason for the majority of people (non-FH general populations) to lower their TC because high TC is actually associated with longevity. Furthermore, increasing P to 10 en% from the current ~7 en% in the USA seems likely to increase health risks unless P is defined with a balance of ω6 and ω3 FAs.

Increased intake of phytosterols (2 g/day) is recommended in the ATPIII issued in 2002. Phytosterols competitively inhibit cholesterol absorption in the intestine, and a transient hypocholesterolemic activity is intended. However, inhibition of cholesterol absorption results in upregulation of cholesterol synthesis as seen with cholestyramine treatment, and a possible associated elevation of isoprenyl intermediates may cause unfavorable side effects as pointed out in chapter 4. Because the effects and side effects of phytosterols as phytoestrogens need further studies, we do not recommend increasing the intake of phytosterols based on the available data.

10.2.

Guidelines from the American College of Physicians for Using Cholesterol Test

Table 23

Guidelines for using serum cholesterol, high-density lipoprotein cholesterol, and triglyceride levels as screening tests for preventing coronary heart disease in adults – American College of Physicians

1	Patients in whom screening for lipoprotein abnormalities is appropriate should have testing with a total cholesterol level
2	In patients who are screened for the primary prevention of coronary heart disease, the total cholesterol level should be measured once; measurement should be repeated periodically if the measured value is near a treatment threshold
3	Screening for total cholesterol levels is not recommended for young men (younger than 35 years of age) or women (younger than 45 years of age) unless the history of physical examination suggests a familial lipoprotein disorder or at least two other characteristics predict a risk for coronary heart disease
4	Screening for total cholesterol levels in the primary prevention of CHD is appropriate but not mandatory for men 35–65 years of age and for women 45–65 years of age
5	Evidence is insufficient to recommend or to discourage screening for the primary prevention of CHD in men and women 65–75 years of age
6	Screening is not recommended for men and women 75 years of age and older
7	All patients with known CHD (history of MI, angina pectoris, other evidence of coronary disease) or history of vascular disease (such as stroke or claudication) are predicted to be at high risk for CHD and should have lipid analysis, including but not limited to measurement of total serum cholesterol levels

Our comments: As a whole, we accept these guidelines as reasonable. We interpret that measuring TC in younger ages (30 years of age) is useful to find FH, and in older ages to find 'falling trends of TC' as a measure of body's changes in unfavorable directions with increasing risk for CHD and all causes deaths.

10.3.
New Directions of the Japan Society for Lipid Nutrition (1997, 2002)

In the President's Summary, decreasing the LA intake to 4.6 en% while keeping ω3 FAs at 1.3 en% was recommended for healthy people. For the prevention of CHD, cancer and allergic hyperreactivity, a dietary ω6/ω3 ratio of 2 was recommended [Okuyama et al., 1996]. In 2002, the Society adopted a proposal that the direction of nutritional recommendations should be changed so as to reduce the intake of LA [Hamazaki and Okuyama, 2003]. Although the Society members have not reached an agreement regarding the recommendable ω6/ω3 ratio, it is our understanding that the current ratio (4–5) should be lowered because high ω6/ω3 ratio-related diseases (cancer and other inflammatory diseases) are increasing in Japan. It is also our understanding that both ALA (PUFA) and HUFA (EPA+DHA) should be included in roughly equal amounts.

10.4.
Workshop of the International Society for the Study of Fatty Acids and Lipids (ISSFAL), NIH, Washington, D.C., 1999

Over 30 specialists from 10 different countries agreed to publish a new direction of lipid nutrition as summarized in table 2. Reduction of LA from current ~7 en% (USA) to about 2 en% was advised, and adequate intakes were set for three major ω3 FAs separately (ALA, EPA and DHA).

Prevention of Coronary Heart Disease

Our comment: The recommended adequate intakes of EPA and DHA were lower than the average intakes of Japanese, but with 2 en% dietary LA, the proportion of ω6 in tissue HUFA would approach that of Japanese. It is our understanding that no unfavorable effects have been noted by the amounts of fish oil ingested by Japanese.

10.5.

Lipid Nutrition for the Prevention of Coronary Heart Disease and Other Late-Onset Diseases – New Directions

Because of metabolic competition between ω6 and ω3 FAs (fig. 70–74), advising to increase the intake of both high LA vegetable oils and fish oil ω3 FAs is not rational. Likely evidence from a follow-up study is reported below.

Fig. 92

Fish intake and risk of incident heart failure – Fried fish vs. non-fried fish

Data taken from Mozaffarian et al. [2005].

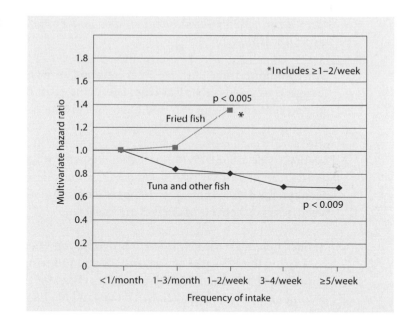

Among 4,738 adults age ≥65 years and free of congestive heart failure (CHF) at baseline in 1989–90, usual dietary intake was assessed using a food-frequency questionnaire. In a participant subsample, consumption of tuna or other broiled or baked fish, but not fried fish, correlated with plasma phospholipid ω3 FAs. Incidence of CHF was prospectively adjudicated.

Results: During 12 years' follow-up, 955 participants developed CHF. In multivariate-adjusted analyses, tuna/other fish consumption was inversely associated with incident CHF, with 20% lower risk with intake 1–2 times/week (hazard ratio [HR] = 0.80), 31% lower risk with intake 3–4 times/week (HR = 0.69), and 32% lower risk with intake ≥5 times/week (HR = 0.68), compared with intake <1 time/month (p trend = 0.009). In similar analyses, fried fish consumption was positively associated with incident CHF (p trend = 0.01). Dietary long-chain ω3 fatty acid intake was also inversely associated with CHF (p trend = 0.009), with 37% lower risk in the highest quintile of intake (HR = 0.73) compared with the lowest.

Conclusion: *Among older adults, consumption of tuna or other broiled or baked fish, but not fried fish, is associated with lower incidence of congestive heart failure. Confirmation in additional studies and evaluation of potential mechanisms is warranted.*

Our comments: The authors' interpretation of possible mechanisms is 'Although frying adds other FAs from the frying oil, it does not reduce the absolute ω3 content. Frying may add oxidation products or, with partially hydrogenated oils, *trans*-FAs, and frying may also have adverse health effects when oils/fats are used repeatedly for frying'. However, lipid peroxides in foods are relatively safe as a diet with increased contents of lipid peroxides suppressed hepatic carcinogenesis in experimental animals [Hirose et al., 1989]. We understand the data to indicate that the intake of LA (ω6) should not be increased when beneficial effects of ω3 FAs are expected.

Background for the new directions in lipid nutrition:
1 'Raising the P/S ratio of dietary FAs and reducing cholesterol intake' was found to be ineffective (and sometimes risky) for the prevention of CHD in long-term intervention trials. The ω6 and ω3 FAs compete at many metabolic processes, and a recommendation 'to increase both the ω6 and ω3 FAs at the same time' is not rational.
2 The TC value and LDL/HDL ratio may be markers for energy balance, intake of different types of FAs and the severity of familial hypercholesterolemic subjects and those with similar genetic disorders (FH). However, sci-

entific evidence is not enough to conclude that high TC and high LDL/HDL ratio are major causes of CHD and other late-onset disease mortalities.

3 No (or little) positive associations have been observed between TC and CHD in general populations in Japan (40 years old), Korean population (30–65 years old) and aged populations in the Western countries (50 years old), in which the proportions of FH are likely to be very low. Importantly, the number of reports has been increasing to indicate that cancer and all-cause mortalities tend to be lower for higher TC groups among general populations, at least up to 280 mg/dl of TC.

4 Evidence has accumulated that 'excessive intake of LA (ω6) and relative ω3 FA deficiency' increases the proportion of ω6 eicosanoid precursors in tissue phospholipids, leading to enhanced inflammatory, thrombotic and arrhythmic events which are proposed to be major mediators for fatal diseases. Some epidemiological studies performed in the USA do not fit the above conclusion, but careful interpretation of the results and extension to lipid nutritional guidelines are required as pointed out in chapter 7.

New directions:

Based on these observations and interpretations, the following new directions in lipid nutrition are recommended for the prevention of CHD and other late-onset diseases.

1 Maintaining the intake of ω3 FAs at or increasing it to the level of average Japanese (~1.3 en%, ALA and EPA+DHA in roughly equal amounts) is recommended.

2 Reducing total fat energy to <25 en% is recommended.

3 Reducing ω6 FAs to 3 en% from 6 to 7 en% in the US and Japan is recommended with recognition that the needed essential amount of LA is <1 en% and that ω3 FAs perform many essential roles.

4 Except for FH and those with related genetic disorders, we do not recommend general populations with TC values below 280 mg/dl to limit their intake of cholesterol or to lower their TC values.

Acknowledgments

This work was supported in part by grants from the Ministry of Health, Labor, and Welfare Japan. We are indebted to Dr. Shinkan Tokudome, Professor and Dr. Teruo Nagaya, Associate Professor, Graduate School of Medicine, Nagoya City University, for their helpful advice in interpreting the statistical data, and Dr. Hama Rokuro for the information on some important epidemiological studies we had overlooked, which we (H.O. and T.H.) obtained as a subscriber of 'The Informed Prescriber', Iyakuhin Chiryo Kenkyukai, Osaka.

Reference Books

Books in English: Lands [2005], Moore [1995], and Ravnskov [2000].
Books in Japanese: The titles are translated from Japanese, and hence may not be completely accurate [Okuyama, 1999; Hamazaki, 2003; Hama, 2004].

References

Adult Treatment Panel III of the National Cholesterol Education Program (NCEP): Bethesda, National Heart, Lung, and Blood Institute, National Institute of Health, NIH Pub No 02-5215, 2002.

AHA/ACC Guidelines for Prevention of Heart Attack and Death in Patients with Atherosclerotic Cardiovascular Disease: 2001 Update.

Albert CM, Campos H, Stampfer MJ, Ridker PM, Manson JE, Willett WC, Ma J: Blood levels of long-chain n-3 fatty acids and the risk of sudden death. N Engl J Med 2002;346:1113–1118.

ALLHAT Officers and Coordinators for the ALLHAT Collaborative Research Group. The Antihypertensive and Lipid-Lowering Treatment to Prevent Heart Attack Trial: Major outcomes in high-risk hypertensive patients randomized to angiotensin-converting enzyme inhibitor or calcium channel blocker vs. diuretic: The Antihypertensive and Lipid-Lowering Treatment to Prevent Heart Attack Trial (ALLHAT). JAMA 2002; 288:2981–2997.

Anderson KM, Castelli WP, Levy D: Cholesterol and mortality: 30 years of follow-up from the Framingham study. JAMA 1987;257:2176–2180.

Ascherio A, Rimm EB, Giovannucci EL, Spiegelman D, Stampfer M, Willett WC: Dietary fat and risk of coronary heart disease in men: cohort follow up study in the United States. BMJ 1996;313:84–90.

Baylin A, Kabagambe EK, Ascherio A, Spiegelman D, Campos H: High 18:2 trans-fatty acids in adipose tissue are associated with increased risk of nonfatal acute myocardial infarction in Costa Rican adults. J Nutr 2003; 133:1186–1191.

Blank C, Neumann MA, Makrides M, Gibson RA: Optimizing DHA levels in piglets by lowering the linoleic acid to alpha-linolenic acid ratio. J Lipid Res 2002;43:1537–1543.

Booth SL, Lichtenstein AH, O'Brien-Morse M, McKeown NM, Wood RJ, Saltzman E, Gundberg CM: Effects of a hydrogenated form of vitamin K on bone formation and resorption. Am J Clin Nutr 2001;74:783–790.

Bosch T, Wendler T: State of the art of low-density lipoprotein apheresis in the year 2003. Ther Apher Dial 2004;8:76–79.

Broughton KS, Wade JW: Total fat and ω3/ω6 ratios influence eicosanoid production in mice. J Nutr 2002;32:88–94.

Burr ML, Fehily AM, Gilbert JF, Rogers S, Holliday RM, Sweetnam PM, Elwood PC, Deadman NM: Effects of changes in fat, fish, and fibre intakes on death and myocardial reinfarction: diet and reinfarction trial (DART). Lancet 1989;2:757–761.

Chan JK, McDonald BE, Gerrard JM, Bruce VM, Weaver BJ, Holub BJ: Effect of dietary alpha-linolenic acid and its ratio to linoleic acid on platelet and plasma fatty acids and thrombogenesis. Lipids 1993;28:811–817.

Christensen JH, Skou HA, Fog L, Hansen V, Vesterlund T, Dyerberg J, Toft E, Schmidt EB: Marine n-3 fatty acids, wine intake, and heart rate variability in patients referred for coronary angiography. Circulation 2001;103: 651–657.

Clarke SD: The multi dimensional regulation of gene expression by fatty acids: polyunsaturated fats as nutrient sensors. Curr Opin Lipidol 2004;15:13–18.

Dawber TR, Nickerson RJ, Brand FN, Pool J: Eggs, serum cholesterol, and coronary heart disease. Am J Clin Nutr 1982;36:617–625.

De Caterina R, Massaro M: Omega-3 fatty acids and the regulation of expression of endothelial pro-atherogenic and pro-inflammatory genes. J Membr Biol 2005;206:103–116.

de Lorgeril M, Renaud S, Mamelle N, Salen P, Martin JL, Monjaud I, Guidollet J, Touboul P, Delaye J: Mediterranean alpha-linolenic acid-rich diet in secondary prevention of coronary heart disease. Lancet 1994;343: 1454–1459.

Dolecek TA, Granditis G: Dietary polyunsaturated fatty acids and mortality in the Multiple Risk Factor Intervention Trial (MRFIT); in Simopoulos AP, et al (eds): Health Effects of ω3 Polyunsaturated FAs in Seafood. World Rev Nutr Diet. Basel, Karger, 1991, vol 66, pp 205 216.

Downs JR, Clearfield M, Weis S, Whitney E, Shapiro DR, Beere PA, Langendorfer A, Stein EA, Kruyer W, Gotto AM Jr: Primary prevention of acute coronary events with lovastatin in men and women with average cholesterol levels: results of AFCAPS/TexCAPS. Air Force/Texas Coronary Atherosclerosis Prevention Study. JAMA 1998;279:1615–1622.

Dwyer JH, Allayee H, Dwyer KM, Fan J, Wu H, Mar R, Lusis AJ, Mehrabian M: Arachidonate 5-lipoxygenase promoter genotype, dietary arachidonic acid, and atherosclerosis. N Engl J Med 2004;350:29–37.

Dyerberg J, Bang HO, Stoffersen E, Moncada S, Vane JR: Eicosapentaenoic acid and prevention of thrombosis and atherosclerosis? Lancet 1978;2:117–119.

Forette B, Tortrat D, Wolmark Y: Cholesterol as risk factor for mortality in elderly women. Lancet 1989;1:868–870.

Fujishima M: What we have learnt from a large-scale epidemiological study in Hisayama Town, Japan. Lipids 2001;12:267–274 (in Japanese).

Gaenzer H, Sturm W, Neumayr G, Kirchmair R, Ebenbichler C, Ritsch A, Foger B, Weiss G, Patsch JR: Pronounced postprandial lipemia impairs endothelium-dependent dilation of the brachial artery in men. Cardiovasc Res 2001;52:509–516.

Geleijnse JM, Giltay EJ, Grobbee DE, Donders AR, Kok FJ: Blood pressure response to fish oil supplementation: metaregression analysis of randomized trials. J Hypertens 2002; 20:1493–1499.

GISSI-Prevenzion Investigators: Gruppo Italiano per lo Studio della Sopravvivenza nell'Infarto miocardico: dietary supplementation with n-3 polyunsaturated fatty acids and vitamin E after myocardial infarction: results of the GISSI-Prevenzione trial. Lancet 1999;354:447–455.

Goodman JE, Bowman ED, Chanock SJ, Alberg AJ, Harris CC: Arachidonate lipoxygenase (ALOX) and cyclooxygenase (COX) polymorphisms and colon cancer risk. Carcinogenesis 2004;25:2467–2472. Epub 2004 Aug 12.

Hama R: Don't Lower Blood Cholesterol and Blood Pressure. Tokyo, Nippon Hyoronsha, 2004 (in Japanese).

Hamazaki T: Relationship between total cholesterol and all cause mortality: Is there a need for lowering total cholesterol in the Japanese population? Curr Topics Nutraceut Res 2004; 2:177–188.

Hamazaki T: Those Who Have High Blood Cholesterol Levels Survive Longer. Tokyo, E-ru Shuppan, 2003 (in Japanese).

Hamazaki T, Okuyama H: The Japan Society for Lipid Nutrition recommends to reduce the intake of linoleic acid; in Simopoulos AP, Cleland LG (eds): The Scientific Evidence. World Rev Nutr Diet. Basel, Karger, 2003, vol 92, pp 109–132.

He K, Merchant A, Rimm EB, Rosner BA, Stampfer MJ, Willett WC, Ascherio A: Dietary fat intake and risk of stroke in male US healthcare professionals: 14 year prospective cohort study. BMJ 2003;327:777–782.

He K, Song Y, Daviglus ML, Liu K, Van Horn L, Dyer AR, Greenland P: Accumulated evidence on fish consumption and coronary heart disease mortality: a meta-analysis of cohort studies. Circulation 2004;109:2705–2711.

Hegsted DM, McGandy RB, Myers ML, Stare FJ: Quantitative effects of dietary fat on serum cholesterol in man. Am J Clin Nutr 1965;17: 281–295.

Hirose M, Kurata Y, Yamada M, Shirai T, Ito N, Ohsawa T: Lack of modifying effects of linolic acid hydroperoxides and their secondary oxidative products on combined 7,12-dimethylbenz[a]anthracene and 1,2-dimethylhydrazine-initiated mammary gland, ear duct and colon carcinogenesis in female Sprague-Dawley rats. Cancer Lett 1989;47:141–147.

Hu FB, Bronner L, Willett WC, Stampfer MJ, Rexrode KM, Albert CM, Hunter D, Manson JE: Fish and omega-3 fatty acid intake and risk of coronary heart disease in women. JAMA 2002;287:1815–1821.

Huang MZ, Watanabe S, Kobayashi T, Nagatsu A, Sakakibara J, Okuyama H: Unusual effects of some vegetable oils on the survival time of stroke-prone spontaneously hypertensive rats. Lipids 1997;32:745–751.

Ingham PW: Hedgehog signaling: a tale of two lipids. Science 2001;294:1879–1881.

Innis SM, Dyer RA: Dietary canola oil alters hematological indices and blood lipids in neonatal piglets fed formula. J Nutr 1999;129:1261–1268.

Irie F, Sairenchi T, Iso H, Shimamoto T: Prediction of mortality from findings of annual health checkups utility for health care programs. Nippon Koshu Eisei Zasshi 2001;48:95–108 (in Japanese with English summary).

Iso H, Kobayashi M, Ishihara J, Sasaki S, Okada K, Kita Y, Kokubo Y, Tsugane S; JPHC Study Group: Intake of fish and n3 fatty acids and risk of coronary heart disease among Japanese: the Japan Public Health Center-Based (JPHC) Study Cohort I. Circulation 2006;17;113:195–202. Epub 2006 Jan 9.

Iso H, Naito Y, Kitamura A, Sato S, Kiyama M, Takayama Y, Iida M, Shimamoto T, Sankai T, Komachi Y: Serum total cholesterol and mortality in a Japanese population. J Clin Epidemiol 1994;47:961–969.

Iso H, Rexrode KM, Stampfer MJ, Manson JE, Colditz GA, Speizer FE, Hennekens CH, Willett WC: Intake of fish and omega-3 fatty acids and risk of stroke in women. JAMA 2001;285:304–312.

Iso H, Sato S, Umemura U, Kudo M, Koike K, Kitamura A, Imano H, Okamura T, Naito Y, Shimamoto T: Linoleic acid, other fatty acids, and the risk of stroke. Stroke 2002;33:2086–2093.

Kang JX, Leaf A: Prevention of fatal cardiac arrhythmias by polyunsaturated fatty acids. Am J Clin Nutr 2000;71(1 suppl):202S–207S.

Kannel WB, Castelli WP, Gordon T: Cholesterol in the prediction of atherosclerotic disease: new perspectives based on the Framingham study. Ann Intern Med 1979;90:85–91.

Kannel WB, Neaton JD, Wentworth D, Thomas HE, Stamler J, Hulley SB, Kjelsberg MO: Overall and coronary heart disease mortality rates in relation to major risk factors in 325,348 men screened for the MRFIT. Multiple Risk Factor Intervention Trial. Am Heart J 1986;112:825–836.

Kendall MJ, Nuttall SL: The heart protection study: statins for all those at risk? J Clin Pharm Ther 2002;27:1–4.

Keys A, Anderson JT, Grande F: Prediction of serum cholesterol responses of man to changes in fats in the diet. Lancet 1957;273:959–966.

Keys A, Mickelsen O, Miller EO, Chapman CB: The relation in man between cholesterol levels in the diet and in the blood. Science 1950;112:79–81.

King DS, Wilburn AJ, Wofford MR, Harrell TK, Lindley BJ, Jones DW: Cognitive impairment associated with atorvastatin and simvastatin. Pharmacotherapy 2003;23:1663–1667.

Kramer JR, Kitazume H, Proudfit WL, Matsuda Y, Williams GW, Sones FM Jr: Progression and regression of coronary atherosclerosis: relation to risk factors. Am Heart J 1983;105:134–144.

Kroman N, Green A: Epidemiological studies in the Upernavik district, Greenland. Acad Med Scand 1980;208:401–406.

Kronmal RA, Cain KC, Ye Z, Omenn GS: Total serum cholesterol levels and mortality risk as a function of age: a report based on the Framingham data. Arch Intern Med 1993;153:1065–1073.

Kuriki K, Nagaya T, Tokudome Y, Imaeda N, Fujiwara N, Sato J, Goto C, Ikeda M, Maki S, Tajima K, Tokudome S: Plasma concentrations of (n-3) highly unsaturated fatty acids are good biomarkers of relative dietary fatty acid intakes: a cross-sectional study. J Nutr 2003;133:3643–3650.

Lands WEM: Fish and Human Health. Orland, Florida, Academic Press, 1986.

Lands WEM: Fish, Omega-3 and Human Health. Champaign, AOCS Press, 2005.

Lands WEM, Libelt B, Morris A, Kramer NC, Prewitt TE, Bowen P, Schmeisser D, Davidson MH, Burns JH: Maintenance of lower proportions of (n-6) eicosanoid precursors in phospholipids of human plasma in response to added dietary (n-3) fatty acids. Biochim Biophys Acta 1992;1180:147–162.

Lopez-Garcia E, Schulze MB, Manson JE, Meigs JB, Albert CM, Rifai N, Willett WC, Hu FB: Consumption of (n-3) fatty acids is related to plasma biomarkers of inflammation and endothelial activation in women. J Nutr 2004;134:1806–1181.

Mabuchi H, Michishita I, Sakai T, Sakai Y, Watanabe A, Wakasugi T, Takeda R: Treatment of homozygous patients with familial hypercholesterolemia by double-filtration plasmapheresis. Atherosclerosis 1986;61:135–140.

Marangoni F, Galli C: Dietary fats of European countries in the Mediterranean area; in Simopoulos A, Visioli F (eds): Mediterranean Diets. World Rev Nutr Diet. Karger, Basel, 2000, 87, pp 78–89.

Matsuzaki M, Kita T, Mabuchi H, Matsuzawa Y, Nakaya N, Oikawa S, Saito Y, Sasaki J, Shimamoto K, Itakura H; J-LIT Study Group. Japan Lipid Intervention Trial: Large scale cohort study of the relationship between serum cholesterol concentration and coronary events with low-dose simvastatin therapy in Japanese patients with hypercholesterolemia. Circ J 2002;66:1087–1095.

McGill HC Jr, McMahan CA, Zieske AW, Sloop GD, Walcott JV, Troxclair DA, Malcom GT, Tracy RE, Oalmann MC, Strong JP: Associations of coronary heart disease risk factors with the intermediate lesion of atherosclerosis in youth. The Pathobiological Determinants of Atherosclerosis in Youth (PDAY) Research Group. Arterioscler Thromb Vasc Biol 2000;20:1998–2004.

Miettinen M, Turpeinen O, Karvonen MJ, Elosuo R, Paavilainen E: Effect of cholesterol-lowering diet on mortality from coronary heart-disease and other causes: a twelve-year clinical trial in men and women. Lancet 1972;2:835–838.

Mizushima S, Moriguchi EH, Ishikawa P, Hekman P, Nara Y, Mimura G, Moriguchi Y, Yamori Y: Fish intake and cardiovascular risk among middle-aged Japanese in Japan and Brazil. J Cardiovasc.Risk 1997;4:191–199.

Moore TJ: Deadly Medicine. New York, Simon & Schuster, 1995.

Mori T, Imaida K, Tamano S, Sano M, Takahashi S, Asamoto M, Takeshita M, Ueda H, Shirai T: Beef tallow, but not perilla or corn oil, promotion of rat prostate and intestinal carcinogenesis by 3,2′-dimethyl-4-aminobiphenyl. Jpn J Cancer Res 2001;92:1026–1033.

Mozaffarian D, Bryson CL, Lemaitre RN, Burke GL, Siscovick DS: Fish intake and risk of incident heart failure. J Am Coll Cardiol 2005; 45:2015–2021.

Multiple Risk Factor Intervention Trial Research Group: Multiple risk factor intervention trial. Risk factor changes and mortality results. JAMA 1982;248:1465–1477.

Nakamura H: Usefulness of lipid-lowering therapy with mevalotin (pravastatin) for the primary prevention aimed at suppressing cardiovascular diseases. 2005. http://www.medical-tribune.jp/congress/mega_study/summary/index.html (in Japanese).

Neaton JD, Wentworth D: Serum cholesterol, blood pressure, cigarette smoking, and death from coronary heart disease. Overall findings and differences by age for 316,099 white men. Multiple Risk Factor Intervention Trial Research Group. Arch Intern Med 1992;152: 56–64.

Newman TB, Hulley SB: Carcinogenicity of lipid-lowering drugs. JAMA 1996;275:55–60.

Ogushi Y: Searching standard values of serum cholesterol for Japanese. Gender Sex Specific Med 2004;2:1221–1229 (in Japanese).

Oh K, Hu FB, Manson JE, Stampfer MJ, Willett WC: Dietary fat intake and risk of coronary heart disease in women. Am J Epidemiol 2005;161:672–679.

Ohara N, Naito Y, Nagata T, Tatematsu K, Fuma SY, Tachibana S, Okuyama H: Exploration for unknown substances in rapeseed oil that shorten survival time of stroke-prone spontaneously hypertensive rats: effects of super critical gas extraction fractions. Food Chem Toxicol 2006;44:952–963[Epub ahead of print].

Okamura T, Kadowaki T, Hayakawa T, Kita Y, Okayama A, Ueshima H: Nippon Data80 Research Group: What cause of mortality can we predict by cholesterol screening in the Japanese general population? J Intern Med 2003;253:169–180.

Okuyama H: Cancers as inflammatory diseases caused by excessive intake of linoleic acid. Environ Mutagen Res 2003;25:147–157 (in Japanese with English summary).

Okuyama H: High n-6 to n-3 ratio of dietary fatty acids rather than serum cholesterol as a major risk factor for coronary heart disease. Eur J Lipid Sci Technol 2001;103:418–422.

Okuyama H: Life Style-Related Diseases, for which Medicines Are Ineffective, Are Prevented by Revolutionary Lipid Nutrition. Nagoya, Reimei Shobou, 1999 (in Japanese).

Okuyama H: Prevention of excessive linoleic acid syndrome. Lipid Technol Newslett 2000;6: 128–132.

Okuyama H, Fujii Y, Ikemoto A: n-6/n-3 Ratio of dietary fatty acids rather than hypercholesterolemia as the major risk factor for atherosclerosis and coronary heart disease. J Health Sci 2000;46:157–177.

Okuyama H, Huang M-Z, Miyazaki M, Watanabe S, Kobayashi T, Nagatsu A, Sakakibara J: Two factors in fats and oils that affect survival time of stroke-prone spontaneously hypertensive rats: n-6/n-3 ratio and minor components; in Riemersma RA, et al (eds): Essential Fatty Acids and Eicosanoids. Champaign, AOCS Press, 1998, pp 273–276.

Okuyama H, Kobayashi T, Watanabe S: Dietary fatty acids – the n-6/n-3 balance and chronic elderly diseases: excess linoleic acid and relative n-3 deficiency syndrome seen in Japan. Prog Lipid Res 1996;35:409–457.

Panagiotakos DB, Chrysohoou C, Pitsavos C, Menotti A, Dontas A, Skoumas J, Stefanadis C, Toutouzas P: Risk factors of stroke mortality: a 40-year follow-up of the Corfu cohort from the Seven-Countries Study. Neuroepidemiology 2003;22:332–338.

Pella D, Thomas N, Tomlinson B, Singh RB: Prevention of coronary artery disease: the south Asian paradox. Lancet 2003;361:79.

Pietinen P, Ascherio A, Korhonen P, Hartman AM, Willett WC, Albanes D, Virtamo J: Intake of fatty acids and risk of coronary heart disease in a cohort of Finnish men. The Alpha-Tocopherol, Beta-Carotene Cancer Prevention Study. Am J Epidemiol 1997;145: 876–887.

Pischon T, Hankinson SE, Hotamisligil GS, Rifai N, Willett WC, Rimm EB: Habitual dietary intake of n-3 and n-6 fatty acids in relation to inflammatory markers among US men and women. Circulation 2003;108:155–60. Epub 2003 Jun 23.

Ratnayake WM, Plouffe L, Hollywood R, L'Abbe MR, Hidiroglou N, Sarwar G, Mueller R: Influence of sources of dietary oils on the life span of stroke-prone spontaneously hypertensive rats. Lipids 2000;35:409–420.

Ravnskov U: The Cholesterol Myths. Washington, New Trend Publishing, 2000.

Ravnskov U: The questionable role of saturated and polyunsaturated fatty acids in cardiovascular disease. J Clin Epidemiol.1998;51: 443–460.

Reilly MP, Pratico D, Delanty N, DiMinno G, Tremoli E, Rader D, Kapoor S, Rokach J, Lawson J, FitzGerald GA: Increased formation of distinct F2 isoprostanes in hypercholesterolemia. Circulation 1998;98:2822–2828.

Rissanen H, Knekt P, Jarvinen R, Salminen I, Hakulinen T: Serum fatty acids and breast cancer incidence. Nutr Cancer 2003;45:168–175.

Ross R: Atherosclerosis: an inflammatory disease. N Engl J Med 1999;340:115–126.

Rubins HB, Robins SJ, Collins D, Fye CL, Anderson JW, Elam MB, Faas FH, Linares E, Schaefer EJ, Schectman G, Wilt TJ, Wittes J: Gemfibrozil for the secondary prevention of coronary heart disease in men with low levels of high-density lipoprotein cholesterol. Veterans Affairs High-Density Lipoprotein Cholesterol Intervention Trial Study Group. N Engl J Med 1999;341:410–418.

Sacks FM, Pfeffer MA, Moye LA, Rouleau JL, Rutherford JD, Cole TG, Brown L, Warnica JW, Arnold JM, Wun CC, Davis BR, Braunwald E: The effect of pravastatin on coronary events after myocardial infarction in patients with average cholesterol levels. Cholesterol and Recurrent Events Trial investigators. N Engl J Med 1996;335:1001–1009.

Sampath H, Ntambi JM: Polyunsaturated fatty acid regulation of gene expression. Nutr Rev 2004;62:333–339.

Sato T, Ozaki R, Kamo S, Hara Y, Konishi S, Isobe Y, Saitoh S, Harada H: The biological activity and tissue distribution of 2′,3″-dihydrophylloquinone in rats. Biochim Biophys Acta 2003;1622:145–150.

Scandinavian Simvastatin Survival Study (4S) Group: Randomised trial of cholesterol lowering in 4444 patients with coronary heart disease: the Scandinavian Simvastatin Survival Study (4S). Lancet 1994;344:1383–1389.

Schatz IJ, Masaki K, Yano K, Chen R, Rodriguez BL, Curb JD: Cholesterol and all-cause mortality in elderly people from the Honolulu Heart Program: a cohort study. Lancet 2001; 358:351–355.

Shekelle RB, Shryock AM, Paul O, Lepper M, Stamler J, Liu S, Raynor WJ Jr: Diet, serum cholesterol, and death from coronary heart disease. The Western Electric study. N Engl J Med 1981;304:65–70.

Shepherd J, Cobbe SM, Ford I, Isles CG, Lorimer AR, MacFarlane PW, McKillop JH, Packard CJ: Prevention of coronary heart disease with pravastatin in men with hypercholesterolemia. West of Scotland Coronary Prevention Study Group. N Engl J Med 1995;333: 1301–1307.

Shibata S, Kumagai S, Watanabe S, Suzuki T, Yasumura S, Suyama Y: Relationship of serum lipids to 10-year deaths from all causes and cancer in Japanese urban dwellers aged 40 years and over. J Epidemiol 1995;5:87–94.

Shields PG, Xu GX, Blot WJ, Fraumeni JF Jr, Trivers GE, Pellizzari ED, Qu YH, Gao YT, Harris CC: Mutagens from heated Chinese and US cooking oils. J Natl Cancer Inst 1995; 87:836–841.

Shirasaki S: Relationship between health examination data and mortality. Are high serum cholesterol and obesity no good for health? Nihon Iji Sinpo 1997;3831:41–48 (in Japanese).

Simon JA, Hodgkins ML, Browner WS, Neuhaus JM, Bernert JT Jr, Hulley SB: Serum fatty acids and the risk of coronary heart disease. Am J Epidemiol 1995;142:469–476.

Simopoulos AP, Leaf A, Salem N Jr: Workshop on the Essentiality of and Recommended Dietary Intakes for Omega-6 and Omega-3 Fatty Acids. J Am Coll Nutr 1999;18:487–489.

Smith SC Jr, Blair SN, Bonow RO, Brass LM, Cerqueira MD, Dracup K, Fuster V, Gotto A, Grundy SM, Miller NH, Jacobs A, Jones D, Krauss RM, Mosca L, Ockene I, Pasternak RC, Pearson T, Pfeffer MA, Starke RD, Taubert KA: AHA/ACC Scientific Statement: AHA/ACC guidelines for preventing heart attack and death in patients with atherosclerotic cardiovascular disease: 2001 update: a statement for healthcare professionals from the American Heart Association and the American College of Cardiology. Circulation 2001;104:1577–1579.

Smith WL: Cyclooxygenases, peroxide tone and the allure of fish oil. Curr Opin Cell Biol 2005;17:174–182.

Song WO, Kerver JM: Nutritional contribution of eggs to American diets. J Am Coll Nutr 2000;19(5 suppl):556S–562S.

Song YM, Sung J, Kim JS: Which cholesterol level is related to the lowest mortality in a population with low mean cholesterol level: a 6.4-year follow-up study of 482,472 Korean men. Am J Epidemiol 2000;151:739–747.

Stemmermann GN, Chyou PH, Kagan A, Nomura AM, Yano K: Serum cholesterol and mortality among Japanese-American men. The Honolulu (Hawaii) Heart Program. Arch Intern Med 1991;151:969–972.

Strandberg TE, Salomaa VV, Naukkarinen VA, Vanhanen HT, Sarna SJ, Miettinen TA: Long-term mortality after 5-year multifactorial primary prevention of cardiovascular diseases in middle-aged men. JAMA 1991; 266:1225–1229.

Strandberg TE, Salomaa VV, Vanhanen HT, Naukkarinen VA, Sarna SJ, Miettinen TA: Mortality in participants and non-participants of a multifactorial prevention study of cardiovascular diseases: a 28 year follow up of the Helsinki Businessmen Study. Br Heart J 1995;74:449–454.

Suzuki S, Oshima S: Influence of blending oils on human serum cholesterol. Eiyogaku Zasshi 1970;28:194–198 (in Japanese).

Suzuki K, Mori M, Nakata A, Kawashimo H, Imai K, Abe N, Takeuchi K, Arai C, Hasegawa M, Takayama Y, Morishita T, Shirai T, Komazawa T: Epidemiological studies on various serum lipids in healthy normal Japanese city habitants (II). Relationship between various serum lipids levels and ischemic ECG changes or aortic pulse wave velocity values for 1984. Domyaku Koka (Atherosclerosis) 1987;15:1547–1556 (in Japanese with English summary).

Tanizaki Y, Kiyohara Y, Kato I, Iwamoto H, Nakayama K, Shinohara N, Arima H, Tanaka K, Ibayashi S, Fujishima M: Incidence and risk factors for subtypes of cerebral infarction in a general population: the Hisayama study. Stroke 2000;31:2616–2622.

Tarui S: Clinical approaches to glycogenoses: analyses on muscle, liver, leucocytes and erythrocytes (review). Nippon Naika Gakkai Zasshi 1987;76:1347–1365 (in Japanese).

Tsuji H, Kitagawa N, Uchida T, Tsuta Y, Sato T, Ikuno H, Morisaki K, Tsuji T, Nanbu I: Higher cholesterol level is associated with lower mortality in a general Japanese population. Osaka Igaku 2004;38:10–15 (in Japanese with English summary).

Ueshima K: Follow-up of the basal examinations of circulatory disease performed in 1980. Nihon Junkanki Kanri Kyogikai Zassi 1997;31: 231–237 (in Japanese).

Ueshima H, Okayama A, Saitoh S, Nakagawa H, Rodriguez B, Sakata K, Okuda N, Choudhury SR, Curb JD, INTERLIPID Research Group: Differences in cardiovascular disease risk factors between Japanese in Japan and Japanese-Americans in Hawaii: the INTERLIPID study. J Hum Hypertens 2003;17:631–639.

Ulmer H, Kelleher C, Diem G, Concin H: Why Eve is not Adam: prospective follow-up in 149650 women and men of cholesterol and other risk factors related to cardiovascular and all-cause mortality. J Womens Health (Larchmt) 2004;13:41–53.

Verschuren WMM, Jacobs DR, Bloemberg BPM, Kromhout D, Menotti A, Aravanis C, Blackburn H, Buzina R, Dontas AS, Fidanza F, Karvonen MJ, Nedelijkovic S, Nissinen A, Toshima H: Serum total cholesterol and long-term coronary heart disease mortality in different cultures: twenty-five-year follow-up of the seven countries study. JAMA 1995;274:131–136.

Wannamethee G, Shaper AG, Whincup PH, Walker M: Low serum total cholesterol concentrations and mortality in middle aged British men. Brit Med J 1995;311:409–413.

Weggemans RM, Zock PL, Katan MB: Dietary cholesterol from eggs increases the ratio of total cholesterol to high-density lipoprotein cholesterol in humans: a meta-analysis. Am J Clin Nutr 2001;73:885–891.

Weverling-Rijnsburger AW, Blauw GJ, Lagaay AM, Knook DL, Meinders AE, Westendorp RG: Total cholesterol and risk of mortality in the oldest old. Lancet 1997;350:1119–1123.

Woodhill JM, Palmer AJ, Leelarthaepin B, McGilchrist C, Blacket RB: Low fat, low cholesterol diet in secondary prevention of coronary heart disease. Adv Exp Med Biol 1978; 109:317–330.

Yam D, Eliraz A, Berry EM: Diet and disease – the Israeli paradox: possible dangers of a high omega-6 polyunsaturated fatty acid diet. Isr J Med Sci 1996;32:1134–1143.

Yamada T, Strong JP, Ishii T, Ueno T, Koyama M, Wagayama H, Shimizu A, Sakai T, Malcom GT, Guzman MA: Atherosclerosis and omega-3 fatty acids in the populations of a fishing village and a farming village in Japan. Atherosclerosis 2000;153:469–481.

Yokoyama M, Origasa H: JELIS Investigators: Effects of eicosapentaenoic acid on cardiovascular events in Japanese patients with hypercholesterolemia: rationale, design, and baseline characteristics of the Japan EPA Lipid Intervention Study (JELIS). Am Heart J 2003;146:613–620, and http://www.mochida.co.jp/dis/jelis/.

Yoshiike N, Tanaka H: What we have learnt from large-scale epidemiological studies performed in Japan. Area-matched Control Study for Japan Lipid Intervention Trial (J-LIT). The Lipid 2001;12:281–289 (in Japanese).

Subject Index

C-reactive protein (CRP)
    statin response 58
    trans-fatty acid intake studies 134, 135

Docosahexaenoic acid (DHA), anti-atherogenesis activity 123–130

Egg intake
    coronary heart disease risk analysis 11
    total cholesterol response 4, 6
Endothelium-dependent dilation (EDD),
    postprandial lipemia effects 122, 123

Familial hypercholesterolemia (FH)
    apheresis therapy 34
    characteristics 20
    confounding of total cholesterol coronary
        heart disease risk studies 25, 27, 30
    lipid metabolism 31–35

Heart rate variability, coronary heart disease
    studies of dietary intervention 76, 77
Hegsted's equation 2
Highly unsaturated fatty acids (HUFA), types
    84–86
Hypertension
    coronary heart disease risks 26, 81
    stroke risks 81

Inflammation
    cancer 147
    injury-inflammation-ischemia hypothesis of
        atherosclerosis 131, 132
    omega–3 fatty acid effects 127–129
    statin effects 58
    trans-fatty acid intake studies 134, 135
International Society for the Study of Fatty
    Acids and Lipids (ISSFAL), dietary intake
    recommendations 154, 155
Intima-media thickness (IMT), coronary heart
    disease studies of dietary intervention 77–80

Japan Society for Lipid Nutrition, dietary intake
    recommendations 154

Leukotrienes, omega–6/omega–3 fatty acid ratio
    effects 116
Linoleic acid (LA)
    cancer risk studies 144–148
    coronary heart disease epidemiology studies
        112, 113, 133
    dietary recommendations 3, 4, 156, 157
    metabolism 106, 112
    myocardial infarction risk studies 10, 15

serum variability factors 111–118
stroke epidemiology studies 107–110
α-Linolenic acid (ALA)
    coronary heart disease protection 8
    dietary recommendations 3, 4
    myocardial infarction risk studies 10
    prostate cancer risks 146
    serum variability factors 111–118
5-Lipoxygenase, alleles and intima-media
    thickness 78, 79

Myocardial infarction (MI), *see also* Atherosclerosis; Coronary heart disease
    linoleic acid intake risks 10, 15
        α-linolenic acid intake effects 10
        omega–3 fatty acid secondary prevention
            studies 88, 89
        trans-fatty acid intake risks 135–137
Nitric oxide (NO), statin response 58
Nonesterified fatty acid (NEFA), vascular injury
    120, 121

Omega–3 fatty acids
    arrhythmia inhibition 125
    atherosclerosis prevention mechanisms
        adhesion molecule inhibition 125, 126
        docosahexaenoic acid 130
        inflammatory response 127–129
        platelet aggregation inhibition 123
        prostaglandin inhibition 124
    coronary heart disease prevention studies
        Greenland 87
        Japan studies 95–101
        MRFIT Study 94
        primary and secondary prevention 86
        secondary prevention 88, 89
        US Nurses' Health Studies 92, 93
        US Physicians' Health Studies 90, 91
    intake recommendations 157
    linoleic acid cascade inhibition 112
Omega–6/omega–3 fatty acid ratio
    coronary heart disease secondary prevention
        compared with statins 102, 103
    overview of imbalance and disease 85
    platelet and plasma phospholipid fatty acid
        composition response 115
    tissue fatty acid composition response 114
    total cholesterol and coronary heart disease
        subgroup analysis 21–23

Phytosterols, intake recommendations 153
Platelet aggregation, omega–3 fatty acid effects
    123